Toxic City

The publisher and the University of California Press Foundation gratefully acknowledge the generous support of the Peter Booth Wiley Endowment Fund in History.

Toxic City

REDEVELOPMENT AND ENVIRONMENTAL
JUSTICE IN SAN FRANCISCO

Lindsey Dillon

UNIVERSITY OF CALIFORNIA PRESS

University of California Press
Oakland, California

© 2024 by Lindsey Dillon

Library of Congress Cataloging-in-Publication Data

Names: Dillon, Lindsey, 1980– author.
Title: Toxic city : redevelopment and environmental justice in San
    Francisco / Lindsey Dillon.
Description: Oakland, California : University of California Press,
    [2024] | Includes bibliographical references and index.
Identifiers: LCCN 2023041738 (print) | LCCN 2023041739 (ebook) |
    ISBN 9780520396210 (cloth) | ISBN 9780520396227 (paperback) |
    ISBN 9780520396234 (epub)
Subjects: LCSH: City planning—California—San Francisco. |
    Environmental justice—California—San Francisco.
Classification: LCC HT168.S2 D555 2024 (print) | LCC HT168.S2 (ebook) |
    DDC 307.1/4160979461—dc23/eng/20231117
LC record available at https://lccn.loc.gov/2023041738
LC ebook record available at https://lccn.loc.gov/2023041739

For Jeanne Dillon and Marie Harrison

Contents

Illustrations

Acknowledgments

Books that take a long time to write accrue many debts. My gratitude first goes to Garden for the Environment and Hunters Point Family, for introducing me to the Double Rock garden and, by extension, to Bayview-Hunters Point. My experiences in the garden became the basis for my dissertation research in the Geography Department at UC Berkeley, where I was fortunate to learn from so many brilliant scholars. Thank you to Richard Walker for taking me on as student, and for the lessons in political economy and California; and thank you to Jason Corburn, Jake Kosek, Don Moore, and Michael Watts for serving on my qualifying exam and dissertation committees, and for asking crucial questions that helped shape an earlier iteration of what became this book. To Christine Trost, Deborah Lustig, David Minkus, and my fellow graduate student colleagues at the Institute for the Study of Societal Issues—you made me a better thinker, writer, and interlocutor. I remember our weekly afternoon meetings in that lovely wood-paneled room fondly. These days I am especially grateful to Don for brightening my day with funny and insightful text messages and for receiving cute pictures of my toddler in return.

My graduate school training—the ways I learned to think and write, the texts I studied, and the political and theoretical commitments that came to inform my scholarship—came as much from the professors who taught classes and mentored me as it did from the equally brilliant graduate students I studied alongside: Javier Arbona, Jenny Baca, Rachel Brahinsky (who also has written extensively about Bayview-Hunters Point, and this book benefited greatly from her scholarship), Erin Collins, Alicia Cowart (now a wonderful cartographer whom I was lucky to work with in creating maps for this book), Shannon Cram, Aaron deGrassi, Jennifer Devine, Ruth Goldstein, Jenny Greenburg, Katy Guimond, Clare Gupta, Juan Herrera, Josh Jelly-Shapiro, Freyja Knapp, Sarah Knuth, Nathan McClintock, Miri Lavi-Neeman, Greta Marchesi, Laura-Anne Minkoff-Zern, Diana Negrín, Adam Romero, John Stehlin, Alexander Tarr, and Jerry Zee.

As I was finishing my dissertation I received an extraordinary gift—a residential research fellowship at the UC Humanities Research Institute at UC Irvine, with a fantastic group of scholars. There, in the Southern California sunshine, we discussed urban ecologies, read each other's work, and ate a lot of good food. It was the perfect place to wrap up graduate school.

I spent two years at UC Davis as a Chancellors Postdoctoral Fellow, under the mentorship of one of the most prominent environmental justice scholars in the country, Julie Sze. Her book, *Noxious New York*, had influenced my graduate studies, and I couldn't believe my good fortune to work with her as a postdoc and eventually to write articles together. During this time, Julie Guthman also mentored me during hikes up Claremont Canyon, in Berkeley. I am thankful for her wisdom and expertise, and for the exercise. Caren Kaplan and others in the critical militarization studies group at UC Davis welcomed me warmly; I am especially grateful to Caren for her encouragement and for modeling what it means to be a feminist scholar and mentor. It was extra special to share an office with my grad student buddy, Javier, and navigate postgraduate life together. Timothy Choy let me sit in on a STS seminar with a group of wonderful graduate students from the anthropology department during this time; I learned so much during those discussions, and I am thankful for Tim's kindness and insights.

I am beyond fortunate to have found a home in the Sociology Department at UC Santa Cruz, with such a vibrant, interdisciplinary group of scholars. To Chris Benner, Julie Bettie, James Doucet-Battle, Hiroshi

Fukurai, Debbie Gould, Miriam Greenberg, Camilla Hawthorne, Naya Jones, Rebecca London, Steve McKay, Jaimie Morse, Juan Pedroza, Jenny Reardon, Alicia Riley, Veronica Terriquez, and especially Hillary Angelo: your brilliance and generosity are a constant source of inspiration. This book benefited greatly from a writing group with Madeleine Fairbairn, Renée Fox, Alma Heckman, Sari Niedzwiecki, and Amanda M. Smith, who read several chapters and provided feedback that strengthened my argument and analysis. Thank you to the UCSC Institute for the Study of Social Transformation for funding a book manuscript workshop in 2020. My deep appreciation goes to David Pellow, Brinda Sarathy, and Jill Harrison for reading and commenting on a (very) early draft of this manuscript (in the early months of the pandemic nonetheless), and to Miriam Greenberg for chairing the event. Jill in particular has been a mentor, source of inspiration, and friend for many years—this book is better because of her support and feedback.

In my first year as an assistant professor at UC Santa Cruz, Donald Trump was elected U.S. president. A few days later, a colleague, Nick Shapiro, sent an email to a number of environmentally focused scholars and friends, asking: How should we respond? Long email chains moved onto Slack channels, and we eventually named ourselves the Environmental Data and Governance Initiative (EDGI). EDGI provided a way to do *something* during those years—it was satisfying work, often fun, sometimes difficult, and though it took me away from writing this book, that time was entirely worth it. I got to collaborate with amazing scholars across the country, on weekly Zoom meetings (this was before the pandemic, when we thought Zoom was cool). Thank you to Nick, of course, and to Phil Brown, Gretchen Gehrke, Sarah Lamdan, Rebecca Lave, M. Murphy, Matt Price, Chris Sellers, Sara Wylie, Lourdes Vera, and many others. When I returned to working on this book it was with a renewed appreciation for what a thoughtful group of people can accomplish together.

At UC Press I worked with three wonderful editors: Stacy Eisenstark, Chloe Layman, and Naja Pulliam Collins. UC Press lets reviewers opt to share their names, and so I know it was Nicole Fabricant and Juan De Lara who provided crucial, critical, and supportive commentary and improved this book immensely. I admire both of them so much, as thinkers and writers—indeed, their books were models for me in writing my own.

Steve Hiatt commented on an early draft of the manuscript, when I was still searching for a throughline, and Megan Pugh made a tremendous impact on the manuscript, helping me rethink the book's narrative as I was working on the manuscript in its later stages. I am extremely grateful to both of them for their critical insights, and especially to Megan for her keen eye and elegant ways of thinking about a book's structure and argument. My gratitude also goes to Sharon Langworthy for copyediting at the end.

Archivists and librarians at the Bancroft Library, the San Francisco branch of the National Archives, and the History Center at the San Francisco Public Library provided invaluable assistance while researching this book. Although I never met him in person, Alex Wellerstein did the incredibly generous thing of sending me a trove of documents on the Naval Radiological Defense Laboratory, after I had sent him an email query about his work. Those documents inform this book and other writing I've done on the lab. I am grateful to Bradley Angel for his decades-long dedication to environmental justice issues, and also for pulling out boxes of Greenaction's archives one afternoon and answering my questions as he was busy fielding phone calls on Greenaction's many campaigns.

I am so grateful to the many Bayview-Hunters Point residents who shared their stories and perspectives with me. Thank you especially to the organizers and volunteers at Greenaction for Health and Environmental Justice and the Quesada Gardens Initiative, both organizations I volunteered with, and to staff members and volunteers with Literacy for Environmental Justice, who took me for walks in Heron's Head Park and invited me to join class field trips. I am also thankful to the many people I spoke with at state agencies and other local nonprofits, whose perspectives enriched this book. And I am deeply appreciative of Dr. Ahimsa Porter Sumchai for reading and providing comments on portions of the book.

Family and friends loved and grounded throughout this time, especially my uncles Richard and Fred; Dell and Dennis, for their love; Dana, for her support from my grad student application to my tenure letter; Jan, with her expertise in all things San Francisco; my brother Michael, and Lana, Myles, and Weston; and my dad, Mark, one of my best friends. My mom, Jeanne, passed away while I was finishing my dissertation. She had lived with cancer for six and a half years—bravely, often with humor, always with gratitude for each day she had. My mom was a voracious reader, my cross-word puzzle buddy, and usually the reigning family Scrabble champion. I still have her encouraging comments on an early draft of my first publication. She gave me those handwritten notes, in her flawless cursive, a few months before she died. Her love buoys me every day.

My dear, loving partner Jamie has had to live with this book for far longer than any academic's partner should. Indeed, he has never known me, or us, without it. He is patient and encouraging and provides excellent feedback. As I struggled to work during the pandemic lockdown, Zoom university, pregnancy, and the first two brilliant, exhausting, and precious years of our son Charlie's life, he keep me afloat, fed and hydrated, and helped me find time to write. And Charlie—your ear-to-ear smile, infectious laugh, and boundless energy light up my days. Now I get to spend even more time with you.

Introduction

Marie Harrison walked slowly to the podium in the grand legislative chamber at San Francisco's City Hall, pulling a wheeled, portable oxygen tank. "This hearing is long overdue," she said to a packed audience. The 2018 hearing before the city's board of supervisors centered on recent findings that a firm contracted by the U.S. Navy to clean up the Hunters Point Naval Shipyard, Tetra Tech, had falsified data on some of its soil samples. The former military base was contaminated with over half a century of industrial and military waste, including the by-products of a Cold War–era radiological laboratory, and it abutted Bayview-Hunters Point, a mixed industrial and residential neighborhood in the southeastern corner of San Francisco where Marie, a Black woman in her sixties, had raised her children and spent much of her life.

Many Black residents in Bayview-Hunters Point had family connections with the shipyard. Military industrialization during World War II had drawn them, or their parents, to the Bay Area to participate in and benefit from the wartime economy. From the military base's opening in 1941 until its closure in 1974, it provided thousands of jobs to local residents and supported neighborhood businesses. In the 1980s, with public revelations about the extent of radioactive waste and other contaminants at the shipyard, residents began to identify the military base as part of a broader landscape of environmental and racial injustice in Bayview-Hunters Point.

A few weeks before the hearing at which Marie spoke, two of Tetra Tech's employees were charged with the data falsifications and sentenced to eight months in prison.[1] Yet many people in attendance at the hearing that afternoon saw the tampering with soil samples not as isolated infractions but as another example of racialized environmental vulnerability in the neighborhood. Moreover, they saw environmental remediation at the shipyard as much more than a technical project of reducing and managing toxic risk.[2] Rather, they sought a more expansive form of social and environmental repair for past harms linked to the military base and its afterlives.

Marie Harrison had a long, established career organizing for housing and environmental justice in southeast San Francisco and was known to neighborhood residents and city officials alike. She was closely involved with community oversight of remediation at the shipyard. Although her health had declined, when she stood at the microphone that afternoon, her words were pointed and precise. Marie demanded a "comprehensive cleanup" of the shipyard, which she qualified as cleanup "not just for the new folks that can buy the new homes." Postremediation redevelopment plans for the old military base include thousands of homes, millions of square feet of office space, and waterfront parks.[3] At the time Marie spoke, several hundred people lived in new townhomes on part of the base, even as the navy's remediation work continued throughout the rest of it. Yet the redevelopment of the shipyard and nearby waterfront properties had generated dust and other particulate matter that, some residents argued, contributed to existing respiratory health problems. They felt disposable in relation to new, high-end residential projects in their neighborhood and exposed to the by-products of redevelopment. Dust and airborne particulates from remediation and new construction added to existing industrial emissions in Bayview-Hunters Point, from a sewage treatment plant, open-air industrial facilities, idling diesel trucks, and two broad freeways that run down the length of the neighborhood. Marie admonished city officials for ignoring the concerns of longtime Bayview-Hunters Point residents. "Listen to people in the neighborhood, they know what is going on. They live it, breathe it, every single day," she told them. For Marie, the falsified soil samples were only the latest environmental injustice in Bayview-Hunters Point. She had spent the 1990s and early 2000s organizing against power plants in the neighborhood and was increasingly concerned with the impacts of environmental cleanup and urban redevelopment on some of San Francisco's poorest residents. "I get angry," Marie told the packed crowd, "when I see a three- or four-year-old with asthma. Is that by design for our community?" Marie articulated a desire for remediation at the shipyard and throughout the Bayview-Hunters Point neighborhood as a form of justice for past and ongoing harms—her demands for environmental repair were part "grievance and grief."[4] Marie ended her remarks by saying, "I'm tired, but I want to be made whole. I have lost way too many people."

Toxic City studies the politics of environmental remediation and urban redevelopment in Bayview-Hunters Point—a neighborhood produced through histories of industrialization, militarization, and state abandonment, and also through the everyday and extraordinary work of neighborhood organizers and residents like Marie to make the neighborhood a better place to live. For more than half a century, Bayview-Hunters Point residents have navigated and resisted the harsh effects of the loss of manufacturing, maritime, and military jobs; racialized neglect by state agencies; industrial pollution; and the toxic legacies of war. Today, the historically Black neighborhood is in the path of gentrifying redevelopment that extends from the city's financial district southward to the Hunters Point Shipyard. Throughout San Francisco's southeastern waterfront, state agencies and private companies are cleaning up and redeveloping former industrial piers, power plants, and other contaminated sites (the five-hundred-acre shipyard is the most notable) as a high-end waterfront that emphasizes consumption, nature, and a particular, market-oriented version of urban sustainability.[5] San Francisco is a case study of post-industrial greening—an urban process most often found in cities across the Global North, where industrial economies have restructured around service, finance, and tech capital.[6] These sectoral shifts have rendered industrial built environments obsolete and available for new development, at the same time that greening and sustainability have become, as sociologist

Miriam Greenberg puts it, "instrumentalized to support broader goals of economic growth."[7]

Large-scale remediation and redevelopment in Bayview-Hunters Point ought to be a socially, economically, and ecologically reparative process. Instead, as Marie alluded to in her speech, remediation and redevelopment are complicated and uneven in their social benefits and have contributed to new forms of dispossession, marginality, and environmental harm. For example, urban restructuring in southeast San Francisco has occurred in tandem with, and has exacerbated, a decline in the city's Black population.[8] In 1970, one in seven San Franciscans, or close to 14 percent of the population, identified as African American. In 2020 that number was 5.4 percent. Since 2010, moreover, most of this population decline has been traceable to Black residents leaving Bayview-Hunters Point.[9] These sociospatial changes raise questions about the place of Black residents in San Francisco's green urban future.

Toxic City also examines the domestic impacts of militarization through the lens of a neighborhood and city negotiating the future of a contaminated military base. Geographer Shiloh Krupar defines militarization broadly as "the social processes by which society composes itself for the production of weapons and national defense."[10] Militarization is also a spatial process; domestically, it builds up cities and local economies, creates new geographies of migration and displacement, and produces toxic landscapes through nuclear and chemical weapons production as well as other harmful by-products of mobilizing for war. Cycles of militarization built up and contributed to the economic unraveling of Bayview-Hunters Point; while the military and the Cold War–era radiological laboratory also left the marginalized neighborhood to the environmental legacies of war and nuclear weapons development. The politics of remediation and redevelopment in Bayview-Hunters Point thus involves negotiating the social and ecological impacts of war. It also broadens scholarly and activist understandings of the scales of environmental injustice in the United States, which are generally domestic in focus, to include geographies of U.S. imperialism.

Bayview-Hunters Point residents have not just been impacted by these dynamics. In the nine years I spent volunteering with organizations, attending meeting and events, interviewing people, and reading through

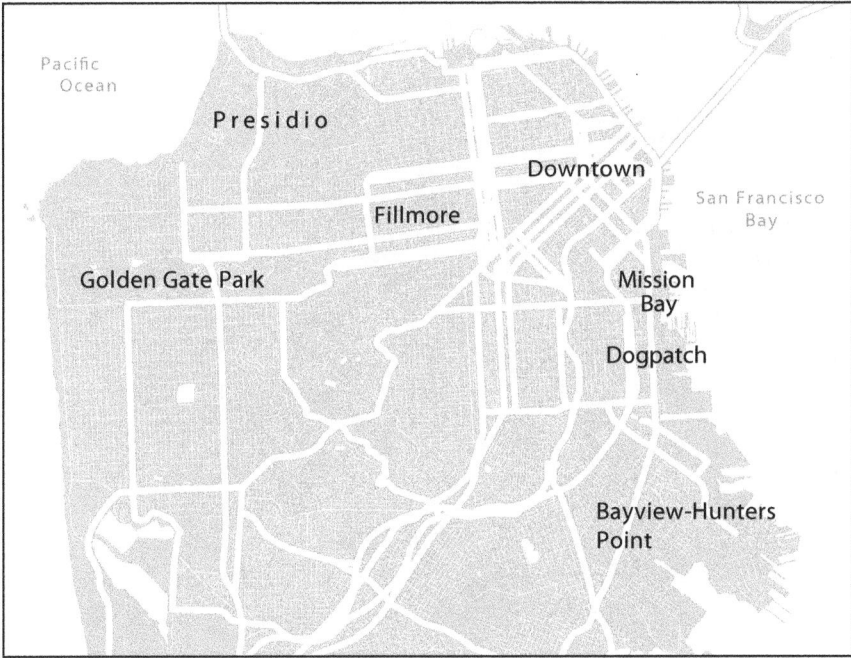

Map 1. Current (2020s) map of select neighborhoods in San Francisco. SOURCE: Created by Alicia Cowart.

archives, I saw the myriad ways residents critiqued and resisted exclusionary urban development and environmental injustice and organized around alternative visions of urban environmental repair. Generations of residents have worked to rebuild dilapidated public housing units and transform weedy, trash-filled lots into gardens. They have chained themselves to the gates of power plants, organized protests and marches, and advanced their own definitions of toxic risk. Their critiques, demands, organizing strategies, and everyday practices, over many decades, have made the neighborhood a better place to live, work, care, and play. In the process, residents have drawn connections between the slow violence of environmental toxicity and histories of racial capitalism, militarization, and uneven development, and have employed multiple tactics—working through, against, and beyond the state—to realize their goals.[11] Their desires for a just remediation at the Hunters Point Shipyard were not about

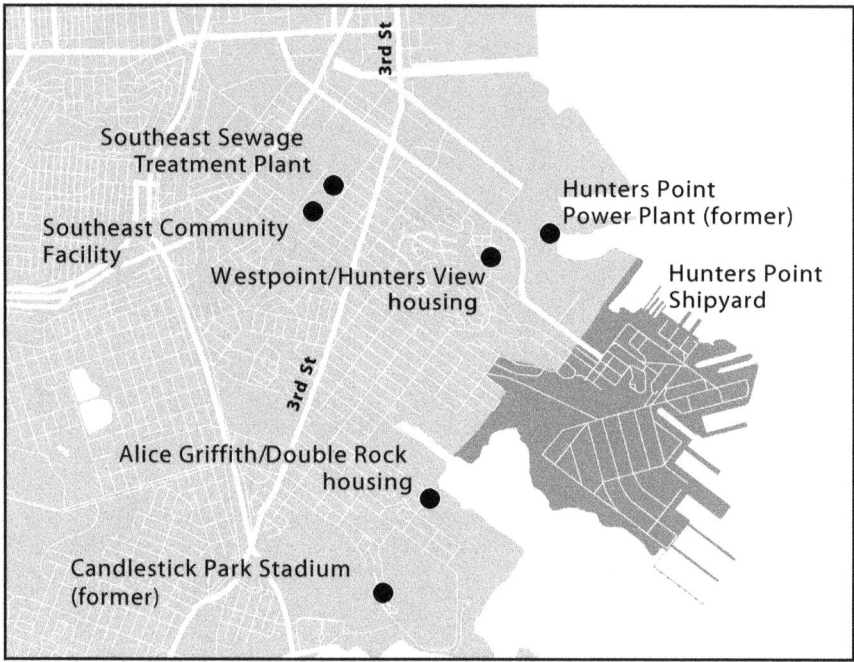

Map 2. Current (2020s) map of Bayview-Hunters Point. SOURCE: Created by Alicia Cowart.

replicating a premilitarized ecology, for example. Rather, they sought acknowledgment of and amends for past and ongoing harms related to the military base, in the context of broader, long-standing goals of creating a better, safer, and healthier future for Bayview-Hunters Point residents, and especially Black residents, many with family histories connected to former slave states in the U.S. South. Organizers like Marie sought a form of urban environmental repair that acknowledged these histories and lived experiences and aimed to build a different city and a different world.

URBAN RESTRUCTURING IN SOUTHEAST SAN FRANCISCO AND MEMORIES OF REDEVELOPMENT

Environmental justice in Bayview-Hunters Point is connected with the history and politics of urban development in San Francisco—in short, with struggles not just to improve a place but with the capacity to stay in place. Marie's concern that cleanup at the shipyard should not simply benefit the "new folks who can buy the new homes," for example, speaks to the backdrop of gentrification in the politics of environmental remediation. Understanding the connection between environmental justice and urban development in Bayview-Hunters Point requires some political-economic, geographic, and historical context.

Since the late nineteenth century, southeast San Francisco has been an industrial, working-class area, where the city effectively pushed most of its noxious, waste-producing industries and racialized, working-class populations. In recent decades, however, state and private investment has remade much of the southeast waterfront, especially areas closer to downtown San Francisco, such as Mission Bay, South Park, SOMA (South of Market), and the Dogpatch. The influx of capital and subsequent reshaping of these mixed industrial and working-class neighborhoods began with the ending of the Cold War in the 1980s and the conversion of research and development industries in Silicon Valley (located about forty-five minutes south of San Francisco, by car) from defense contractors to computer and internet technology (IT) companies. The accumulation of tech capital in Silicon Valley led to the first "dot-com boom" in San Francisco in the second half of the 1990s, as well as to the gentrification of working-class neighborhoods, such as the historically Latinx Mission District as well as SOMA and South Park, along the southeastern waterfront, north of Bayview-Hunters Point.[12] In the 2010s, the success of newer tech companies such as Twitter, Salesforce, Facebook, and Google led to the "tech boom 2.0," which contributed to what anthropologist Manissa Maharawal has called "hyper-gentrification" in the city, especially in the southeast.[13] The tech boom 2.0 also coincided with a construction boom in Mission Bay, beginning with the University of California's new hospital complex and, subsequently, a biotech corridor. (Genentech Hall, University of California, San Francisco's [UCSF]'s first research building in Mission Bay had opened in 2003.)[14] Formerly a landscape of warehouses and railyards, today, in addition to the hospital campus and biotech companies, Mission Bay and the adjacent Dogpatch neighborhood are the location of high-end condominiums, trendy restaurants, and boutique shops selling eclectic, pricey wares. Property values throughout the southeast have also increased as a result of a new municipal light rail line, the

T-Third Street, which began running in 2007. [15] The new rail line traces the length of the southeastern waterfront, from the financial district, downtown, to Bayview-Hunters Point. The light rail line addressed a historic lack of public transportation in the southeast, yet the project met with heavy criticism from Bayview-Hunters Point residents for failing to hire local workers and for disrupting businesses along Third Street, the main commercial corridor of the neighborhood. [16]

The influx of tech capital and the transformation of the industrial, working-class waterfront in southeastern San Francisco also coincided with the U.S. military's decision to transfer the Hunters Point Naval Shipyard back to the city, and thus with the availability of land—a scarce commodity in a city bordered by water on three sides. This too was an outcome of the end of the Cold War. At the same time that defense research firms in Silicon Valley morphed into an IT industry in the 1980s, the military began rethinking its geopolitical strategies, which involved restructuring its physical footprint, both domestically and abroad. Military bases and nuclear production facilities had "mushroomed" across the United States during and after World War II, some of them resembling small cities. [17] During the Cold War, defense spending continued to produce military-industrial regions in New England, the Southwest, and California, especially the San Francisco Bay Area. With the end of the Cold War, the military began the process of shuttering hundreds of bases—now considered "surplus" properties—and transferring the land back to states and municipalities through the Base Realignment and Closure Act (BRAC). [18] The navy announced it would return the Hunters Point Shipyard to the city of San Francisco in 1991, two years after the U.S. Environmental Protection Agency (EPA) designated the military base a Superfund site—a regulatory term for the most contaminated sites in the country. In a move that reflected the city's political-economic shift to a service, financial, and tech-centered economy, in 1999 the city selected Lennar Corp., one of the largest residential real estate developers in the country, to redevelop the shipyard.

Along with the transition of parts of southeast San Francisco from a working-class, industrial waterfront into a gentrified tech hub, private, nonprofit, and state entities have pursued myriad forms of postindustrial greening projects in the area. One of the most ambitious of these in southeast San Francisco is the Blue Greenway—a thirteen-mile network of small greening projects which have transformed, or aim to transform, abandoned contaminated piers and irregular parcels of land into wetlands, small parks, and kayak landings. The Blue Greenway imaginatively assembles these sites into one greenbelt running down the southeast waterfront, connected by a bike lane. The stated goal of the Blue Greenway is to "connect neighbors to their waterfront and serve as a catalyst for community building, employment opportunities, and economic vitality."[19] This rhetoric reflects how urban greening projects are usually carried out, and often received, as well-intentioned public goods.[20] Yet they are also known to increase property values and—in the absence of robust affordable housing policies—contribute to what scholars term "green gentrification."[21] Indeed, the Blue Greenway follows the trajectory of gentrification down Third Street, from China Basin (north of Mission Bay) to Candlestick Point State Park, next to the Hunters Point Shipyard.

Thus, the city's partnership with Lennar to redevelop the shipyard, coinciding with corporate and state investment in Mission Bay, the new transportation route, the reimagining of the industrial waterfront as a site of nature and consumption, and the edgy, hip, yet upscale street culture and aesthetic emerging in the southeastern neighborhoods, has led to an increased interest and development pressure—in part by real estate speculators, but also by households priced out of other parts of the city—in Bayview-Hunters Point.[22]

Contributing to this new interest in Bayview-Hunters Point real estate, in 2006 the San Francisco Redevelopment Agency (SFRA) designated the neighborhood a "Redevelopment Project Area." Because of the SFRA's history in both the Fillmore/Western Addition (a neighborhood that I will refer to simply as the Fillmore) and Bayview-Hunters Point, many residents had complex feelings about the SFRA's project, with some residents opposing it entirely. In the late 1950s, the SFRA began evicting tenants and bulldozing entire streets in the Fillmore as part of a large urban renewal project. At the time, the Fillmore was the center of Black cultural and political life in the city.[23] To this day, the SFRA's project in the Fillmore—which took property by eminent domain, evicted thousands of Black households, and shuttered Black businesses—exemplifies the phrase popularized by the writer James Baldwin, that urban renewal meant "Negro removal."[24]

I heard this critique of redevelopment from longtime resident and organizer, Patricia.[25] Patricia was in her late seventies when I first met her, at a navy-led shipyard remediation meeting in 2011. A widely respected community figure with a long career organizing in Bayview-Hunters Point, Patricia was a regular at the public meetings on both redevelopment and remediation in Bayview-Hunters Point that I attended during the 2010s. Dressed smartly, often in a colorful blazer and a hat, she would deliver eloquent critiques to city and navy officials, sometimes referencing events from thirty or forty years ago as a reminder that the contemporary politics of cleanup and redevelopment were part of a longer history of Black struggle in Bayview-Hunters Point. I introduced myself to Patricia on a chilly weekday evening in January, after the participants of that month's remediation meeting had spilled out of the Bayview YMCA and stood huddled on the sidewalk, saying their goodbyes. She seemed pleased, though unsurprised, that a graduate student was interested in environmental politics in Bayview-Hunters Point. She gave me her card and told me to call her in a few days. I did, and we spoke for several hours. Patricia's family had moved to San Francisco from Texas in the 1940s, and she lived "at the door of the shipyard as a kid." Her organizing career began with the welfare rights movement in the 1960s and expanded over the course of her life to include housing, redevelopment, and in the 1990s, remediation. She was instrumental in establishing the Hunters Point Shipyard Restoration Advisory Board (RAB), which provided a forum for community oversight of the shipyard remediation process. As we spoke, she described how the neighborhood had changed over the course of her life, which brought her to changes currently afoot. She did not trust "redevelopment," which she interpreted as a form of anti-Black racism. "Look at Hunters View," she told me, referring to the redevelopment of a public housing development—historically, with a majority of Black residents. The public housing development, which was also known as West Point, had been unlivable for decades, and its residents were sorely in need of decent housing. Yet Patricia feared that the public-private redevelopment project—a denser, mixed-income, mixed-ownership housing development; half of the units would be sold at market rate prices—would replicate what had happened in the Fillmore. "It's to get people out," Patricia told me. She clarified, "Redevelopment comes to get people." Patricia's take on redevelopment and the way she experienced it as linked with an effort, on the part of city agencies, to remove longtime Black residents from the neighborhood was not unique. "They're fixing things up, but it's not for us," one Black resident told the *San Francisco Chronicle* in 2008.[26] His comment reflected widespread concerns that the myriad redevelopment projects taking place in Bayview-Hunters Point were not intended to benefit the longtime Black community.

Patricia's was not the only perspective in Bayview-Hunters Point at that time, nor was it the only perspective among Black Bayview-Hunters Point residents. For other residents I spoke with in the 2010s, the SFRA's promise of jobs and urban improvements represented a welcome and belated endeavor. Yet enough residents feared the impacts of SFRA-led redevelopment that a group called Defend Bayview-Hunters Point circulated a petition to put the SFRA's designation to a referendum on the 2006 city ballot. The petition gathered enough signatures for inclusion on the ballot, but a court ultimately rejected the referendum due to a technicality (the petition voters signed did not contain the full text of the original sixty-two-page policy, which was the subject of the referendum). On September 27, 2006, the anniversary of the 1966 Hunters Point uprising, Defend Bayview-Hunters Point organized a march down Third Street—the main commercial corridor in the neighborhood—to protest the rejected ballot measure and the SFRA redevelopment plan. Demonstrators carried a wide banner that read, "We Shall Not Be Moved! 33,000 signatures demand to be counted."[27]

A few weeks before I spoke with Patricia I had met Louis, one of the organizers of Defend Bayview-Hunters Point.[28] To get to Louis's home office on Third Street, I boarded the T-Third Street rail line at Embarcadero Station, downtown. The T-Third initially follows the curve of the city's waterfront, but after crossing Mission Creek, it hops onto Third Street for the remainder of its southbound route. When it reaches Bayview-Hunters Point, Third Street morphs into the commercial heart of the neighborhood. Along Third Street, in the 2010s, you could find Auntie April's Chicken, Waffles & Soul Food and the Ruth Williams Memorial Theater, named for an important civil rights activist from Bayview-Hunters Point.

Third Street in Bayview-Hunters Point is both a commercial corridor and a boundary, a line between Hunters Point, to the east of Third, and the

Bayview district, to the west. Until World War II, Hunters Point was largely characterized by cattle yards, shipyards, and factories. Federal spending for wartime workers in the 1940s transformed Hunters Point into a neighborhood within a few short years, while in the postwar era, the San Francisco Housing Authority built so many units there that by the 1960s the neighborhood had the largest concentration of public housing in the city.[29] Bayview, to the west of Third Street, is an older residential neighborhood, with more single-family houses. Before World War II, the Bayview district was primarily home to working-class, white ethnic communities. In the 1950s Black households started to leave public housing in Hunters Point and rent or purchase homes in the Bayview, and the two neighborhoods were increasingly seen as one. The new name, Bayview-Hunters Point, reflected an expansion of a Black sense of place in southeast San Francisco, even as the two neighborhoods are still seen as distinct places.

Toward the end of our interview, Louis brought me to a small outdoor porch just off the kitchen of his second-story flat to show me clusters of beautiful red strawberries, growing in wooden barrels. As I admired the berries, a large, coffin-shaped black box resting against a wooden fence on the grass below took me by surprise. Louis explained that it was a prop used during an antiredevelopment rally a few years before. For Louis, the SFRA's designation of the neighborhood as a Redevelopment Project Area was undemocratic. "The only other city that has bigger land grab is New Orleans, and at least they had a flood," he told me, referencing the entrepreneurial and exclusionary rebuilding of historically Black neighborhoods in New Orleans that followed Hurricane Katrina in 2006. The SFRA redevelopment project in Bayview-Hunters Point was not the same as the large, federally funded, and multisectoral remaking of New Orleans, which included the privatization of public schools.[30] Nor was it the same as the SFRA's urban renewal project in the Fillmore in the mid-twentieth century. The agency had no plans to evict tenants and raze whole city streets in Bayview-Hunters Point, for example, nor did it plan to take property through eminent domain. Still, for Louis these events were cut from the same cloth. Each involved urban land acquisition by outside actors and represented a form of accumulation by dispossession, predicated on or at least connected to the displacement of Black residents.[31] And yet, in spite of the provocative symbolism of the coffin—suggesting the death of Bayview-Hunters Point as a

Black neighborhood—Louis was hopeful about the future. I asked where he saw the neighborhood in thirty years and he told me, "I think you'll see an increase in the Black population. That's why I'm telling people, look, whoever owns Third Street owns Hunters Point." He was emphasizing the importance of local, Black-owned businesses as a bulwark against the growing pressures of gentrification.

Louis had been deeply involved in protesting SFRA's redevelopment project in Bayview-Hunters Point, and he didn't think much of Lennar's massive housing project at the shipyard either. At one point during our conversation he waved in the direction of the shipyard. "Are you going to build something over there," he asked, rhetorically, "or are you going to take care of these people over here [on Third Street] with some of that money?" A few moments later, he added, "I mean, you're not doing anything but trying to run the people off. And you're going to build on land that's not clean?" Louis had attended shipyard RAB meetings in the 2000s, along with Patricia and Marie. Although he was uninterested in Lennar's redevelopment project on the shipyard, he was involved in community oversight of the cleanup process. As I discuss in chapter 3, many residents felt strongly about toxic cleanup at the shipyard and saw cleanup as a form of environmental justice. At the same time, cleanup was connected with all the uncertainties of urban redevelopment in Bayview-Hunters Point and the question of who gets to benefit from the postindustrial, postmilitary transformation of the neighborhood.

SEEKING ENVIRONMENTAL JUSTICE THROUGH, AGAINST, AND BEYOND THE STATE

Most scholars locate the origins of the U.S. environmental justice movement in a campaign against the siting of a hazardous waste landfill in a poor, majority Black area of Warren County, North Carolina, that began in the late 1970s. The threat of the hazardous waste landfill brought together a coalition of white and Black residents, who initially developed legal and technical arguments against locating the landfill in Warren County. When recourses to state institutions failed, residents shifted tactics. They reached out to local civil rights activists and Black church leaders, who contacted nationally influential organizations, including the United Church of Christ (UCC) Commission on Racial Justice and the Southern Christian Leadership Conference. This institutional support and involvement of key figures from the civil rights movement elevated the struggle in Warren County from a local to a national issue. The evolving opposition to the hazardous waste landfill also coalesced around the argument that race and racism were underlying factors in the selection of Warren County as the landfill site. The new coalition adopted the civil disobedience and direct action tactics of the civil rights movement, such as lying down in the middle of the road to prevent trucks with hazardous waste from arriving in town.[32]

Although the hazardous waste landfill was eventually built in Warren County, the merging of the civil rights and antitoxics movements led the chairman of the Congressional Black Caucus, Walter Fauntroy (who had participated in the antilandfill demonstrations), to commission a report by the U.S. General Accounting Office (GAO) "to determine the location of hazardous waste landfills and the racial and economic status of surrounding communities" in the U.S. Southeast region.[33] The GAO report, published in 1983, indeed found a correlation between geographies of race and toxic waste, with three out of four hazardous waste landfills in the southeastern U.S. located in majority Black, poor areas. The GAO report was followed four years later by the UCC Commission on Racial Justice's landmark study, *Toxic Wastes and Race in the United States: A National Report on the Racial and Socio-Economic Characteristics of Communities with Hazardous Waste Sites*. The study, published in 1987, surveyed major U.S. metropolitan areas and found race to be the most significant variable in predicting the location of commercial hazardous waste facilities, and that three out of five Black and Latinx Americans lived in communities with uncontrolled toxic waste sites. The UCC report bolstered the GAO's conclusion, as well as the arguments of protestors in Warren County, that Black and other minoritized communities in the United States were overwhelmingly exposed to industrial emissions, hazardous waste landfills, and other forms of toxic waste.

These empirical studies and the growing adoption of environmental justice as an organizing framework among a diverse range of long-standing social movements coalesced with the National People of Color Environmental Leadership Summit in 1991, held in Washington, D.C. The summit included hundreds of delegates from across the United States, including Puerto Rico and the Marshall Islands, and produced a document titled *The Principles of Environmental Justice*. These principles included "the fundamental right to political, economic, cultural, and environmental self-determination of all peoples"; "the right to participate as equal partners at every level of decision-making, including needs assessment, planning, implementation, enforcement and evaluation"; "the rights of victims of environmental injustice to receive full compensation and reparations for damages as well as quality health care"; and an opposition to "military occupation, repression, and exploitation of lands, peoples and cultures, and other life forms."[34] The document advanced different political strategies and orientations to the state, reflecting the diversity of social movement actors at the event.

Recent scholarship offers a critical assessment of the environmental justice movement in the decades since the 1991 Summit. Geographers Laura Pulido and Juan De Lara, for example, lament that some of the radical political imaginaries articulated in the *Principles* "have slowly been replaced by more moderate appeals to the liberal state for inclusion and redress."[35] Yet, they argue, the state is not equipped to address the fundamental roots of racialized environmental inequalities in racial capitalism and colonialism. Indeed, as many scholars, including Pulido, David Pellow, Malini Ranganathan, and Julie Sze have argued, the state is part of the production and perpetuation of environmental injustices.[36] The water crisis in Flint, Michigan, which national media outlets began reporting on in 2014, is one example of Pulido's argument that environmental racism is not simply the "disproportionate exposure" of Black, Indigenous, and people of color to hazardous waste, as it is commonly defined in the academic and policy literature, but a form of state-sanctioned violence.[37] Indeed, even as forty years of environmental justice scholarship has clearly detailed racial environmental inequalities as systemic problems, and even as the U.S. government mandates federal agencies to consider environmental justice in their programs and policies, environmental justice lawsuits regularly fail in the courts, environmental justice Title VI complaints are rejected to a degree that is "truly breathtaking in scope," and government enforcement of environmental laws is consistently discriminatory.[38] Moreover, as sociologist and geographer Jill Harrison demonstrates, within the EPA itself environmental justice is often seen as a performative "box to check" rather than as fundamental to the agency's work.[39] As David Pellow puts it, the current social order, including state power, "stands as a fundamental obstacle to social and environmental justice."[40] Consequently, as Pulido, Ellen Kohl, and Nicole-Marie Cotton argue, the continued reliance on or engagement with state institutions by environmental justice organizations "has inhibited their ability to achieve their goals."[41] Instead of working in partnership with state agencies, they argue, the environmental justice movement "needs to see the state as an adversary and directly challenge it."[42] In this vein, Pulido and De Lara argue for a return to the political horizons articulated by the 1991 Summit, with the goal, more broadly, of connecting environmental justice organizing with abolitionist and decolonial projects and imaginaries. Writing about practices of abolitionist environmental justice, scholar Ki'Amber Thompson explains that "abolition pushes us to build solutions outside the state and critique its ability to provide justice and healing."[43] And as scholars Dina Gilio-Whitaker and Jaskiran Dhillon argue, environmental justice organizing needs to grapple with the state as a *settler* state and take seriously Indigenous claims to sovereignty.[44] These analytical frameworks and practices offer essential, necessary critiques of political strategies that expect or depend on state institutions to deliver justice.

The struggles and experiences of Bayview-Hunters Point residents in this book reinforce a concept of the state as creating, sanctioning, and reproducing racialized environmental vulnerabilities and harms. And yet, while I believe it is important to analyze and critique social movement tactics and strategies (with the goal of building a more just world), this book takes a different focus. What I offer, rather, is a historical, place-based study of how activists and everyday people relied on multiple strategies, both radical and reformist, sometimes working with or through, and sometimes working in opposition to or beyond, state institutions.[45] Aspects of this book thus reinforce Harrison's assertion that environmental justice activism has always employed a diversity of tactics, as well as sociologist Tracy Perkins's study of environmental justice activism in California, which examines how activists navigate "a state that is both a cause of their problems and, at times, at least a partial solution to them."[46] Additionally, the environmental justice claims discussed in this book do not necessarily come from self-identified activists or from nonprofit organizations, but are also, in some cases more simply, the ideas, critiques, and experiences that I heard articulated on porch steps, in public meetings, or in the neighborhood newspaper—perspectives, in other words, that do not always rise to the analytical level of a "movement," even as they indicate the potential for radical change. In a sense, there are similarities between the place-based, multigenerational critical perspectives discussed in this book and aspects of what geographer Clyde Woods called a *blues epistemology*: an interpretive tradition of theory and practice, a working-class, Black American vision of a "non-oppressive society" that has been "daily constructed . . . through a variety of cultural practices, institution-building activities, and social movements."[47] Thus, while the book emphasizes the state's role in environmental injustice in Bayview-Hunters Point, I am not critical of the ways residents have sought justice from the state or through state spaces, such as in public hearings or advisory committees. Rather, I see these spaces as sites where complex perspectives and demands on the state have been expressed and where important gains had been made. Indeed, one of the things that struck me over the years that I researched this book was the nuanced critiques residents brought to public meetings and hearings, such as Marie Harrison's testimony to the board of supervisors in 2018, where this introductory chapter began.

A final note: the current debate about radical versus reformist versions of environmental justice activism has not yet substantively engaged with the problem of technoscience. As historian M. Murphy and others have pointed out, state environmental regulation relies on a problematic epistemology of chemicals as "discrete entities, as isolated molecules, often represented through abstract structural diagrams," as opposed to "complex bundles of extensive relations."[48] This restrictive understanding of chemicals contributes to an imperceptibility of the immense "chemical relations" in which we are embedded (albeit unevenly) and often has the effect of supporting the industrial-corporate status quo, which continues to pollute poor and vulnerable neighborhoods, not to mention the planet. Environmental justice activists often rely on these same technoscientific habits and disciplines (such as toxicology) in their attempts to prove harms, change policy, and protect vulnerable communities. In doing so, however, they reinforce ways of knowing chemicals and ecologies that arguably undermine the movement's larger goals. In response, scholars have challenged those "modes of politics that rest on dominant evidentiary representations of harm" and have explored "less celebrated modes of activism to collectively argue for multiple concepts of toxic politics and reproductive justice in a permanently polluted world."[49] This perspective and focus is important, but I want to linger on Murphy's point that "it is difficult to talk about chemicals in any other way."[50] Just as Bayview-Hunters Point residents grappled with the limitations of state institutions even as they pushed the state to do better, so they also grappled with the limitations of these technoscientific discourses and practices in complex ways. For example, residents have labored to collect data and "prove" environmental harms in their neighborhood and to speak the language of state regulation. And they have demonstrated how these epistemological frameworks fail their community and reinforce a toxic status quo. The point I want to make is that technoscientific categories and the evidentiary requirements of regulatory agencies add another layer to the double bind many activists face in relation to the state and to the question of what constitutes radical environmental justice politics.

METHODS AND APPROACH

In many ways this book project began in 2004, when—as part of a weekend gardening class I took with an organization called Garden for the Environment—I started volunteering at a community garden in the Double Rock public housing development (also known as Alice Griffith), near the Hunters Point Naval Shipyard. The Double Rock garden originated in 1995 as a project of the San Francisco League of Urban Gardeners (SLUG). SLUG did not introduce gardening to Bayview-Hunters Point; indeed, a 1988 article in its newsletter notes that one of the "pluses of working in Hunters Point are the large numbers of people who know gardening and farming."[51] This was likely because many Bayview-Hunters Point residents or their parents had moved to San Francisco from rural southern states during the second half of the twentieth century, bringing with them agricultural knowledge and expertise. But in 2003 SLUG was suspended from receiving municipal contracts, its main revenue source, after the organization's director inappropriately used public funds to influence San Francisco's mayoral election. SLUG folded, and the Double Rock garden went to seed.[52]

This was part of the history of the garden when, the following year, I began spending Saturday mornings helping another local organization, Hunters Point Family, rehabilitate the garden—cleaning up trash, laying mulch for new fruit trees, and planting rows of vegetables. To get to the garden, I drove, with a few other volunteers, down Third Street, before turning left, past warehouses and down a quiet road, into the housing project. It was a bumpy drive. Much of Third Street was then a dusty, unpaved construction zone, as the city was in the process of building the T-Third rail line. The SFRA redevelopment project for the broader neighborhood was being contested and raising concerns about gentrification. I could, in a literal sense, feel the unevenness of redevelopment, as the car jostled along the road. It was unclear, at the time, how redevelopment would improve the lives of people living in the dilapidated Double Rock housing units, some of which faced the garden. I could sense the uncertainty that swirled in the air along with the construction dust unearthed from the bowels of Third Street. While working at Double Rock I also learned about the city's plan to clean up and redevelop the Hunters Point Shipyard. Indeed, I could see some of the military's old buildings and structures from the garden. This weekend volunteer work—weeding next to an abandoned military base and seeing the upheaval along Third Street—raised questions for me about the political economy and ecology of urban redevelopment and about place-based articulations of environmental justice.

Methodologically, this book is informed by participant observation, interviews, and archival work conducted between 2010 and 2019. Between 2011 and 2013 I attended monthly navy-led meetings on remediation at the shipyard and occasional meetings on various aspects of redevelopment. For eight months between 2011 and 2012, I volunteered on Saturday mornings with Quesada Gardens Initiative in the Bayview, an organization that turns vacant, often trash-filled lots into gardens and other green spaces. I also joined weekend volunteer days at two different reconstructed wetlands in Hunters Point, just north of the shipyard. From 2015 to 2016 I volunteered with Greenaction for Environmental and Health Justice, helping with their community mapping and Environmental Justice Task Force project, which included backend data entry, taking notes at task force meetings, and helping with community outreach. During all these years I participated in environmental justice demonstrations at the shipyard and at state environmental agency headquarters throughout the city and attended other cultural and political events in Bayview-Hunters Point. I conducted sixty-seven semistructured interviews (including follow-up interviews) with longtime residents, nonprofit workers, city planners, and environmental scientists (none of these categories are exclusive), along with numerous informal conversations in gardens, on porch steps, at demonstrations, and while socializing at meetings. I also conducted archival work at the San Francisco Public Library, UC Berkeley's Bancroft Library, and the San Francisco branch of the National Archives.

As I finalize this manuscript in the spring of 2023, the company redeveloping the Hunters Point Naval Shipyard, the navy, Tetra Tech, new homeowners on the former military base, hundreds of San Francisco police officers (the city police department had leased a building on the shipyard), and Bayview-Hunters Point residents are entangled in a web of lawsuits revolving around contamination, cleanup, and toxic exposure at the shipyard. In April, an environmental nonprofit, Public Employees for Environmental Responsibility, filed an official complaint with the navy's inspector general contending that the navy had consistently downplayed toxic risks at the shipyard and misled the public about potential harms.[53] This book represents an analysis of an ongoing story. Still, what I find important are the ways Bayview-Hunters Point residents have, over many years, advocated for forms of urban environmental repair and redevelopment that centered their ideas of racial and environmental justice.

Although I focus on the relationship between the shipyard and Black Bayview-Hunters Point residents, I want to make clear that southeast San Francisco is today one of the most racially diverse areas of the city. As working-class households are pushed out of other parts of the city, and as Black residents continue to leave, or are pushed out of, San Francisco, Bayview-Hunters Point now has a significant number of Latinx, Asian, Pacific Islander, and, increasingly, white households.[54] Concerns about redevelopment, about toxic cleanup, and about environmental justice are by no means confined to the neighborhood's Black community. Still, Black residents have lived, organized, and shaped the built and social environment of Bayview-Hunters Point for over eighty years. Some of the people involved in these issues today are following in the footsteps of their parents and grandparents. Many are deeply invested in ensuring that remediation and redevelopment acknowledge what they see as historical injuries and that these projects center the well-being of the neighborhood's Black community.

I have chosen to focus on their stories because, first, the historical racialization of Hunters Point, and later, Bayview-Hunters Point, as a Black neighborhood, along with the tradition of political activism by Black residents, is central to understanding the critiques, contestations, and justice claims regarding remediation and redevelopment today. Relatedly, many Black Bayview-Hunters Point residents have personal connections to the Hunters Point Shipyard, through family members who worked there during and after World War II. Over the course of my fieldwork I heard longtime Black residents talk about the shipyard with a complex mixture of pride in their national service, or of their family members' service, alongside a refusal of the environmental toxicity and harm the military base later came to represent. Moreover, most, though not all, residents who showed up at the naval environmental remediation meetings I attended were Black residents, indicating the significance of the remediation process to the longtime Black community.

This book is shaped by my social positionality as a white woman and as an outsider to the neighborhood. In many ways Bayview-Hunters Point has been overstudied by white policy makers and academics, albeit from a deficit, if not dehumanizing, framework. Throughout the course of my research I read countless policy reports and news articles from the 1960s to the present that rehearsed a "culture of poverty" analysis and portrayed residents, at best, as helpless victims of larger forces.[55] Many longtime residents have, for good reason, skepticism toward researchers not from the neighborhood, particularly white people like myself who live elsewhere (I commuted to San Francisco from the East Bay, and later from Santa Cruz, while researching this book). Thus, for all the stories and perspectives people shared with me, and which inform this book, there was likely just as much they declined to say to me, for personal and historical reasons. As I researched and wrote and rewrote this book, I struggled with the historically extractive ways of knowing that still predominate in the social sciences. This book cannot escape that history. However, I have tried to ground my analysis in the voices and analytical perspectives of neighborhood residents I met over the years. These voices and perspectives are not representative of everyone who lives or lived in Bayview-Hunters Point. Even within the longtime Black community, there is quite a diversity of opinions on redevelopment and what the future of the neighborhood should look like, not to mention the best way to get to that future.

When I interviewed Bernice, a longtime resident and organizer, in 2012, she told me about her wariness of outside nonprofits.[56] She said: "People still come to the Bayview with their ideas to spend money, to use their statistics. . . . I say to them, I don't know how you got that money, but this is what is important to us. So if you want to work with us, this is what you have to work on." This book reflects my attempt to work on issues to which Bernice and many others dedicated their lives. One of the benefits of having worked on this book over many years is that I have had that much time to listen, to follow up, to question and change my own thinking and ask better questions, and to try to write something that "stands with" the Bayview-Hunters Point residents I have long admired.[57] This book is necessarily imperfect, but I hope, at the very least, readers will come away with some of that admiration, too.

MAP OF THE BOOK

Chapter 1, "The Wastelanding of Southeast San Francisco," traces land use patterns in San Francisco from the era of Spanish and Mexican colonialism to the mid-twentieth century, showing how geographies of racism and toxic waste developed in relation to political-economic shifts and colonial regimes over time. By the second half of the twentieth century, zoning regulations, World War II militarization, municipal housing policies, and urban planning projects had produced Hunters Point (and later, Bayview-Hunters Point) as a Black neighborhood confronted with multiple forms of state abandonment.

Chapter 2 focuses on the 1960s and 1970s, when Black residents of Bayview-Hunters Point organized against oppressive state practices and for community control over the provision of housing, health care, and food through what I call Black *counterplanning*. Counterplanners sought to build a livable neighborhood in Bayview-Hunters Point, by and for Black San Franciscans. I argue that these efforts reveal a tradition of urban planning and environmental repair informed by movements for Black self-determination and a critique of the racial state.

The second half of the book shifts to the more recent past, beginning in the 1990s, to explore the politics of remediation and redevelopment at the Hunters Point Shipyard. Chapter 3 details how Bayview-Hunters Point residents have pushed back against the state's technocratic, risk-based remediation project at the shipyard and sought to broaden the military's responsibilities to the community. Rather than accept a circumscribed project of reducing and managing toxic risk, residents have demanded a form of remediation that includes acknowledgment and accountability for past and present harms linked to the military base and its afterlives, including the ways militarization has intersected with racial geographies in San Francisco.

Chapter 4 looks at how residents have protested a seemingly insignificant by-product of new construction along the Hunters Point waterfront: dust. While the shipyard development company and state environmental agencies portray redevelopment dust as a technical issue that can be fixed with better air quality management plans, many residents experience redevelopment dust as a form of environmental racism and part of a longer history of state abandonment, respiratory distress, and Black marginality in San Francisco.

The conclusion, "Reparative Environmental Justice," considers how the ideas of environmental justice can be part of a broader project of reparations for slavery and its afterlives, with reparations conceived of as a "forward-looking target" of building a more just world.[58]

1 The Wastelanding of Southeast San Francisco

In the 1880s, Chinese fishermen operated hundreds of fishing nets staked out across the San Francisco Bay and lived in numerous small camps located on and around Hunters Point. The *San Francisco Chronicle* sent a journalist to report on one of these encampments in 1889. His article described the difficulties of his trek: "From the drawbridge at Fourth and Townsend streets down to Hunters point [*sic*] itself, is a perpetual recurrence of boggy roads, clouds of dust, reeking malarious acres of black mud and stinks that baffle comparison or description."[1] Some of those offensive odors likely came from Butchertown, San Francisco's meatpacking district; in the 1860s, the city's sprawling cattle industry had moved to the tidelands of Islais Creek basin, just north of Hunters Point. Nearby sugar refineries, iron foundries, chemical and petroleum factories, glue and soap factories, tanneries, and shipyards would have added to the atmosphere.[2] When the journalist arrived at Hunters Point, he reported on Chinese fishermen from a distance, observing "dark-blue figures flitting among the hulls or on the docks." Hunters Point, though technically part of the city, was, to the journalist, "an out of the way corner—which seems to have nothing whatever to do with San Francisco."

By 1889, San Francisco had become the financial capital of the U.S. West, and its bustling commercial district and tall buildings were easily recognizable landscapes of urbanism. The journalist saw the underdeveloped, muddy landscape of southeast San Francisco as altogether separate from the part of the city where he likely began his journey, downtown—a separateness that also derived from the presence of Chinese fishermen. In fact, the wealth accumulating downtown relied on the perceived disposability of southeast San Francisco, as a place to locate sprawling, noxious-waste-producing industries. What's more, even as a surge of populist white nativism in San Francisco and across the country had coalesced into an organized movement of anti-Chinese racism, leading to the federal Chinese Exclusion Act of 1882, the economy of the city—and indeed, of the American West—relied on Chinese labor.[3]

Historian Traci Brynne Voyles writes that capitalism depends on spaces of extraction and disposal, and therefore on the discursive rendering of some landscapes as waste-able. She calls the social production of pollutable, marginal landscapes *wastelanding*.[4] Voyles emphasizes that race and racism have historically functioned as signifiers of wastelands—think of the toxic ecologies of uranium mining in Diné, or Navajo, territory, or the hazardous waste dump built in Warren County, North Carolina.[5] Wastelanding, according to Voyles, takes two primary forms: the assumption that racialized lands are valueless, or valuable only for what can be extracted from them, and the material devastation of those places by extractive and polluting industries. Although Voyles does not use the term *racial* capitalism, her concept of wastelanding aligns with Ruth Wilson Gilmore's and Laura Pulido's writings on the ways the production of surplus value depends on the racialized disposability of peoples and places.[6]

In this chapter I explore the wastelanding of southeast San Francisco and the social production of Bayview-Hunters Point. Broadly, I am interested in the historical geographies of racism and waste in the city and how these processes articulate in the making and meaning of a place.[7] To that end, the chapter provides an overview of land use changes and urban development in southeast San Francisco, beginning with the Spanish colonial era, where, I argue, the history of Bayview-Hunters Point as a marginal, "out of the way corner" begins. I follow the example of geographer Louise

Johnson, who writes that any account of urban land "must recognize the significance of its prior occupancy and revisit the colonial past to re-tell some of the histories of its initial dispossession."[8] This chapter reveals how forms of Black political activism—explored in subsequent chapters—are part of the "layered histories embedded in place," including the ways San Francisco urban geographies were produced through multiple colonial projects and imperial routes.[9] This historical geography brings us to the post–World War II decades, when another writer, James Baldwin, visited Hunters Point and announced, "This is the San Francisco America pretends does not exist."[10] By then, 97 percent of Hunters Point residents were Black, and thousands of people lived uphill from a power plant and in close proximity to the military shipyard. This chapter shows how geographies of racism and toxic waste developed in relation to political economic shifts and colonial regimes over time. It also establishes some of the historical and geographical conditions for the emergence of environmental justice activism in 1990s as well as the return of capital investment to southeast San Francisco, in the form of large-scale redevelopment and gentrification, in the 2000s.

A VERY OLD WORLD

When European colonists sailed into what became known as the San Francisco Bay in the eighteenth century, they found a landscape filled with people. According to historian Matthew Booker, the San Francisco Bay was a "crowded and cultivated land" much like California itself, which was, at the time of Spanish colonialization, the most densely populated region of North America.[11] In eighteenth-century California, ecologist M. Kat Anderson writes, "excluding desert and high-elevation areas, it was almost impossible for early Euro-American explorers to go more than a few miles without encountering indigenous people." Anderson concludes, "The New World is in fact a very old world."[12]

The novelist and philosopher Sylvia Wynter suggests that European colonizers perceived the Americas as "uninhabitable torrid zones" in relation to the "habitable temperate zone of Europe."[13] This boundary line between habitable and uninhabitable, derived from medieval Christian geographies, also marked a caesura in conceptions of the human and justified European expropriation of Indigenous land. In contrast to this racial-spatial imaginary, Native Americans in California, as anthropologist Kent Lightfoot and his coauthors detail, were active land managers "who constructed productive anthropogenic landscapes through a variety of methods, including tillage, pruning, seed broadcasting, weeding, selected burns, and irrigation."[14] The Ohlone people of the land today known as the San Francisco Bay Area also relied on tidal marshes along the bay shore, including the wetlands around today's Bayview-Hunters Point. Settler accounts from the time and later anthropological (and mostly still settler) studies reported a large number of Indigenous fisheries and hundreds of shellmounds along the San Francisco estuary. A 1906 survey found sixteen shellmounds in San Francisco proper, and eight of those shellmounds in Hunters Point—indicating the economic and cultural significance of the place.[15]

Spanish ships, en route to the Philippines, occasionally stopped along the California coastline in the seventeenth century, but it was not until the second half of the eighteenth century that Spain colonized the coastal region, including what became the San Francisco Bay Area. By then, expanded networks of trade; marine mammal hunting; and missionary and scientific exploration by Britain, Russia, and the United States threatened Spanish military dominance in the Pacific as well as Spain's inland empire and mining operations in Central and South America.[16] At the time, Spanish mineral extraction in contemporary Mexico, Peru, and Bolivia, which relied on Indigenous and African labor, supplied the coffers of the Spanish crown and was contributing to growth of capitalism in Europe.[17] As a bulwark against the colonial ambitions of other European nations, Spain built a network of Franciscan missions, military presidios, and several small pueblos along the California coastline, from Baja California to the northern regions of the San Francisco Bay.[18]

Spanish colonizers killed many Indigenous Americans and coerced others onto missions, where they provided the backbone of the mission economy. In addition to European diseases, military violence, and forced relocation, the intensive agricultural and ranching operations of the mission economy altered California's coastal landscape, a process historian Alfred Crosby calls "ecological imperialism."[19] Imported animals grazed on the peripheries of mission settlements, trampling on grasslands, while Spanish agriculture introduced new plant species, which quickly spread throughout the region. Meanwhile, Spain prohibited the long-standing Indigenous practice of using fire to manage landscape ecologies through controlled burns, while "the onslaught of alien weeds, free-range cattle, sheep, and pigs, and changes in local hydrology due to irrigation systems disrupted local ecosystems that were the livelihood of California Indians."[20] Indigenous Americans did not passively accede to Spanish colonial intrusion: many fought back, while thousands fled from Spanish mission settlements, sometimes to inland California, where they established new villages or joined other tribes.[21] Meanwhile, Indigenous people living on missions also committed "small acts of resistance" against Spanish colonizers, "such as claiming to be sick, expressing confusion over instructions, or breaking agricultural tools."[22]

Still, Spanish colonialism transformed the political and ecological landscape of the coast, with the effect of establishing land use patterns that influenced subsequent Mexican and U.S. settler colonialism. Between 1776 and 1823, Spain built five missions and a military presidio in what is today the San Francisco Bay Area. Mission San Francisco de Asís was constructed in 1776.[23] During the period of Spanish colonial dominance, the area south of Mission San Francisco de Asís was used as cattle pasture, including part of today's Bayview-Hunters Point neighborhood.[24] In this way, Spanish colonial land uses established that particular site as a kind of hinterland—a place on the edge of the mission settlement, yet central to the mission economy.

These land use patterns, including the relationship between center and periphery, continued through Mexican and U.S. rule. Mexican independence from Spain in 1821 initiated a political-economic transformation in California that was also a spatial and ecological transformation. In contrast to Spanish mercantilism, Mexico opened coastal California to international markets, integrating the region with globalizing racial capitalism. Foreign companies, particularly New England–based tanneries, could now access the region's growing cattle industry, and the San Francisco Bay became further entrenched as an entrepôt on Pacific Rim trading routes.[25] Mexican rule also initiated a new regime of private property. The 1824 Colonization Act authorized the new nation-state to grant large tracts of land for private ownership. The subsequent Secularization Act of 1833 privatized former Spanish mission lands, which were considered prime grazing areas. According to historian Steven Hackel, "Taken together, these laws ushered in the greatest transfer of land and resources in California since the Spaniards set foot in the region. As a result, by 1840 the private rancho had replaced the mission as the dominant social and economic institution in California."[26] As with Spanish missions, the Mexican rancho economy depended on Indigenous labor. Historian Benjamin Madley writes: "To wrangle and process the bovine multitudes ranging over their vast domains—which routinely exceeded 10,000 acres and contained thousands of animals—ranchers turned to California Indians, who served as cowboys, broncobusters, butchers, skinners, tanners, and tallow renders."[27] Indigenous Californians survived European colonialism by fleeing and fighting, and also became laborers within the region's emerging agrarian capitalism.

In 1839 the Mexican citizen José Cornelius Bernal obtained a land grant that included the former pasture of Mission San Francisco de Asís. North of Bernal's land grant was the Mexican trading village of Yerba Buena, which in the following decade U.S. settlers would rename San Francisco. Bernal's property included Islais Creek basin as well as the rocky promontory that Spanish colonizers called Punta de la concha (Point of the shell) and which, under subsequent U.S. occupation, was renamed Hunters Point. Bernal named his land tract Rancho rincón de las salinas y potrero viejo (Corner of salt marshes and the old grazing land), and continued to use the area around Islais Creek for his cattle.[28]

Seven years later, in 1846, the United States declared war on Mexico, in what was considered, even to critics at the time, an imperial war of aggression. Mexico conceded to the United States in 1848. Through the Treaty of Guadalupe Hidalgo, the United States took half of Mexican state territory, including Alta California. That same year, gold seekers from around the world flooded the newly renamed city of San Francisco, as part of the California Gold Rush. Explosive population growth and capital generated by extractive gold and silver mining in the Sierra Nevada built up the small trading village into a cosmopolitan port city and the financial center of the U.S. West.[29]

As part of the 1848 treaty, the United States agreed to honor Mexican land grants if those grants were deemed legal, by U.S. courts, under Mexican property law. In practice, however, land grant court cases often

dispossessed Mexican property holders of their title, transferring ownership to U.S. settlers and speculators. Mexican land grant holders also divided and sold off parts of their property to pay for legal bills incurred in
the process of proving ownership in the U.S. court system, and in other
instances to benefit from a new, speculative real estate market.[30] Bernal
died in 1850, but in 1852 his heirs were forced to prove ownership of their
land before the U.S. Board of Commissioners of California Land Claims.[31]
The board of commissioners ultimately granted Bernal's heirs the right
to their property, but it was a smaller parcel than what Bernal had first
acquired: the historical record shows that he had already sold part of his
land grant—the promontory that became known as Hunters Point—to
real estate developers in 1849, a year before he died.[32] In subsequent years
his family members subdivided other parts of their land grant and sold
these parcels to U.S. homestead and railroad associations; in part, the historical records suggests, to pay off debts incurred during the court case.[33]

The new property owners of Hunters Point had short-lived plans to
subdivide the promontory and build a suburban extension of San Francisco. However, the difficulty of traveling from the city's central business
district to Hunters Point (as exemplified by the journalist's trek four decades later) dampened residential property interest and led the shareholders of large tracts to seek other land uses, especially as parts of the
southeast waterfront were increasingly used for industrial meatpacking,
chemical factories, and other noxious industries. The South San Francisco Homestead and Railroad Association, for example, reorganized as
the South San Francisco Dock Company and marketed parcels on Hunters
Point for industrial rather than residential development.[34] What had been
an economically and culturally important place for the Ohlone people
soon became an industrial district, and ultimately a waste-able landscape,
for U.S. settlers.

U.S. INDUSTRIALIZATION IN SOUTHEAST
SAN FRANCISCO

In the second half of the nineteenth century, urban development in
San Francisco concentrated on the northeastern waterfront. Tideland

reclamation projects reshaped that part of the shoreline and established a commercial waterfront, while the city became a bustling, recognizable landscape of urbanism. Southeast San Francisco, on the other hand, was still dominated by marshes and mudflats that swelled seasonally with the tides.[35] Yet it too was urbanizing and becoming entangled within geographies of global racial capitalism and U.S. imperialism.

On his way to Hunters Point from downtown San Francisco in 1889, the *Chronicle*'s journalist might have seen the Pacific Rolling Mills factory, which supplied iron bars and rods for San Francisco's growing transportation system as well as the Southern Pacific Railroad Company.[36] Near Pacific Rolling Mills was Union Iron Works, owned by Pittsburgh-based Charles Schwab's Bethlehem Steel Company, which manufactured machinery for the Comstock Lode, the famed silver mine in the Sierra Nevada. Union Iron Works also held contracts with both the U.S. and Japanese militaries to build gun boats, cruisers, and battleships.[37] Nearby, Claus Spreckels's sugar refinery imported raw sugar cane from Hawai'i, where it had been cultivated by Chinese, Japanese, and Filipinx labor.[38] Butchertown, located around Islais Creek basin—part of the Spanish mission's former pasture—had become the largest meatpacking district west of Chicago.[39] In 1867 some of California's most prosperous capitalists funded the California Dry Dock on Hunters Point (after several changes in ownership and the construction of additional drydocks, this became the Hunters Point Shipyard), including William Ralston, who owned the Comstock Lode; Isaac Friedlander, who controlled much of the state's overseas wheat trade; and Lloyd Tevis, the president of San Francisco–based Wells Fargo, who made his fortune as a venture capitalist in Ralston's Comstock Lode.[40] In the aftermath of the Spanish–American War and U.S. colonialization in the Philippines, the navy identified the strategic importance of the San Francisco Bay Area to military interests. For example, in 1905 Bethlehem Steel purchased the drydocks at Hunters Point, and the company maintained contracts with the U.S. Navy to build and repair navy ships.[41]

Until World War II, Hunters Point was primarily the location of cattle yards, shipyards, and factories, along with scattered homes and a small but productive Chinese shrimping industry. Residential districts emerged haphazardly in the southeast, alongside the area's industrial and agricultural land uses, and primarily in the adjacent Bayview district. Industrial workers often lived in clusters of company housing tracts near the factories where they worked. For example, in 1885, when Claus Spreckels's sugar refinery expanded its operations, it also built tenement houses and a school nearby.[42] In 1912 the Western Pacific Harbor Tract Company announced plans to build houses in the Bayview, "where some of the thousands employed in the Union Iron Works, Risdon Irons Works, sugar factories, and other works may find homes . . . in easy walking distance of these industries."[43] In addition to these company housing tracts, Chinese, Italian, Maltese, and Portuguese immigrants established small farms in the Bayview and supplied the city's produce market.[44]

This haphazard mixture of working-class residential, industrial, and agricultural land uses was codified by the city's first comprehensive zoning map in 1921. As a tool of city planning—then a nascent professional occupation—zoning regulates land use, specifying where industrial activities can and cannot operate. In theory, zoning protects public health by separating residential from industrial land uses. In practice, zoning laws have historically aimed to protect the health and property values of wealthy city dwellers and commercial business by separating them from manufacturing activities and from manufacturing workers—many of whom, in early twentieth-century cities, were immigrants.[45] These kinds of exclusionary zoning laws left working-class, often racialized, city dwellers vulnerable to the health effects of industrialization. Over time, zoning laws have thus contributed to the sedimentation of racial capitalism in urban landscapes and bodies. The 1921 San Francisco zoning ordinance created an irregular map of "residential," "industrial," and "unrestricted" zones in southeast San Francisco, further entrenching the tight patchwork of industrial and working-class (initially, white ethnic and Chinese) residential land uses in this part of the city, which persists to the present day. This proximity of land uses, undeveloped infrastructure, and low-income immigrant population was also the stated basis for the Homeowner Loan Corporation's (HOLC) designation of much of Bayview and Hunters Point as "hazardous" for investment. Between 1935-1940, HOLC produced maps of U.S. cities with at least forty thousand residents, rating city blocks according to perceived financial risk. These ratings were based largely on demographics and the quality of housing stock, with areas rated "D", or "hazardous" for investment, color-coded red and almost always designating Black neighborhoods, leading to what is today considered the racially discriminatory practice of *redlining*.[46] Redlining in southeast San Francisco preceded the migration of Black wartime workers to the area, yet the classification still would have denied southeast residents federally backed mortgages and other loans for property improvements until the Fair Housing Act of 1968.[47]

So far I have shown how Spanish and Mexican colonial land use patterns influenced urban and industrial development in San Francisco. The Spanish mission's former cattle pasture became the U.S. city of San Francisco's outer industrial zone. The mixture of racialized working-class housing and industrial activities that developed in southeast San Francisco in the late nineteenth century was reinforced through early twentieth-century zoning regulations. By the early twentieth century, then, political economies and racial-spatial imaginaries had produced southeast San Francisco as a waste-able space—a place for noxious industries and disposable populations. These land use patterns established the conditions for the militarization of Hunters Point during World War II, including the development of Hunters Point itself as a residential neighborhood, to service the war effort.

THE MILITARIZATION OF HUNTERS POINT

It was a brisk February afternoon in 2011 that I met Louis at his home office on Third Street in Bayview-Hunters Point. Louis was born in Shelby County, Texas, on the border of Louisiana. He grew up on his family's farm but went to high school in Houston when his father went to work at the shipyards in Orange, Texas, during World War II. "Everyone worked at the shipyard," he remembered. "You either came to California, or you went to the shipyard in Texas." Louis moved to San Francisco in 1950, following his brother, who had served in the war. He worked as a rigger, loading and unloading heavy objects on and off ships. "But I fell in love with [San Francisco] when I was about seven years old," he told me. "I read about it, and then would go down to the movies—back then we paid nine cents to go to the movies—and of course we had to sit way up in the back, and all the whites sitting down below. But they showed San Francisco, and the bridge, and I said, that's where I'm going."[48]

Prior to World War II, the Black population in San Francisco was relatively small compared to much larger Black communities in Oakland and

Los Angeles. This was primarily due to the strength of the white labor movement in San Francisco, which excluded Black San Franciscans from the manufacturing or construction trades. W. E. B. Du Bois observed this when he traveled to California in 1913 as part of a West Coast tour. Du Bois wrote that "the opportunity of the San Francisco Negro to earn a living is very difficult," since "the white trade unions have held the Negro out and down."[49] Nineteenth- and early twentieth-century Black San Franciscans worked predominantly in nonunionized service and domestic jobs; they were virtually excluded from industrial labor in the city until World War II.[50]

Beginning in 1940, federal defense dollars flooded the San Francisco Bay Area, funding sprawling shipyard complexes and new or expanded military installations and attracting workers such as Louis's brother to the West Coast.[51] The navy took over the commercial dry docks in Hunters Point in 1941 and built out the new military base by hundreds of acres, using earth from the surrounding hillsides. Industrial war production in the Bay Area drew in workers from across the country, who migrated to jobs at shipyards in Richmond, Vallejo, Marin City, West Oakland, and Hunters Point or followed family members who did. By 1943, nearly 80 percent of workers employed in heavy industry in the Bay Area worked at the shipyards.[52] War mobilization also shifted the geography of the Great Migration (the large-scale movement of rural Black southerners to the industrial North between 1910–1970) from routes headed to northeastern and midwestern cities to West Coast wartime boomtowns like San Francisco. Between 1940 and 1950 alone, San Francisco's Black population grew by 665 percent, from 4,846 to 43,460 people. Of the Black migrants that arrived in San Francisco between 1940 and 1944, over nine-tenths were from the South and "virtually all" from Texas, Louisiana, and Oklahoma.[53]

As with other industrial sectors, wartime labor in the Bay Area was racially structured, with Black workers relegated to the lowest-paid and most dangerous jobs. The explosion at Port Chicago is a clear example of the violence of this racial division of labor. On the night of July 17, 1944, an ammunition ship at the Port Chicago Army Depot, on the northeastern side of the San Francisco Bay, exploded. The blast killed 320 men and wounded 390 others. Port Chicago was a segregated "Jim Crow base," with white commissioned officers overseeing Black enlisted men as they carried out the dangerous work of handling ammunition. Because of this racial division of labor, the majority of those killed and wounded by the explosion were Black men.[54] Yet Black workers also resisted these labor conditions. When white officers commanded workers at Port Chicago to begin loading ammunition again, soon after the deadly blast, hundreds refused, in protest.[55] A few years earlier, in December 1942, Black sailors at Mare Island Naval Station, near Port Chicago, marched from the military base into the nearby Bay Area town of Vallejo in a "deliberate stand against white vigilantism and official segregation."[56] And, in response to the practice of forcing Black labor into separate, "auxiliary" unions, Black shipyard workers in the Bay Area organized to desegregate shipyard unions, using themes from the nationally popular Double V campaign in their organizing material. Launched in 1942 by the *Pittsburgh Courier*, a Black-owned newspaper with a national circulation, the Double V campaign stood for the wartime struggle waged by Black workers on "two fronts," against fascism abroad and racism at home.[57] According to a letter from a woman in Texas, published in the *Courier*, the Double V campaign was important "because many Americans are more dangerous to us than some of our enemies abroad"—a point reinforced by the deadly explosion at Port Chicago.[58]

Black workers in San Francisco labored for the war effort during the day and came home to segregated neighborhoods at night. Most wartime newcomers moved to the Fillmore district, near downtown San Francisco, which had become the center of Black cultural and political life in the city in the 1920s. By 1930 half of the city's Black population lived in that neighborhood. The Fillmore was also home to a large Japanese community, which was forcibly relocated to internment camps beginning in 1942, after the United States entered the war.[59]

The Fillmore became the primary destination for Black newcomers during World War II, in part because of the sizable Black community who lived there before the war, in part because of rental vacancies created by Japanese internment, and in large part due to housing discrimination—in the 1940s, Black migrants could find few other places in the city to live. Racial housing practices in San Francisco during the war were documented by the well-known Black sociologist Charles S. Johnson. Johnson had been hired by a consortium of local and national civic organizations

To All Shipyard Workers:–

MAKE SUNDAY, MAY 28th,
YOUR D-DAY

(Destroy Discrimination Day)

DON'T FAIL TO ATTEND THIS

Mass Meeting

Hear Attorney George R. Andersen report on the Moore Dry Dock

and Richmond Shipyards Cases

Canal War Apartments Auditorium
Wine and Cutting Streets
Richmond, California

Sunday, May 28th, 2:00 P.M.

HELP PUT THE COURT FIGHT OVER THE TOP

Shipyard Workers Committee Against Discrimination

3

Figure 1. Meeting announcement posted by Shipyard Workers Committee Against Discrimination, May 28, [1943]. SOURCE: Meiklejohn Civil Liberties Institute Collection, Bancroft Library, UC Berkeley.

to research the social conditions of Black wartime migrants to the city.[60] In his study, *The Negro War Worker in San Francisco*, published in 1944, Johnson found that local merchant associations and neighborhood improvement clubs had "hurriedly taken drastic steps to encourage the formation of restrictive agreements and other legal and non-legal devices" to prohibit Black newcomers from moving into white neighborhoods. Moreover, it was "fairly evident" "that a concerted effort is being made . . . to restrict the area of living for Negro families to the present boundaries of the Filmore [*sic*] district."[61] Rents in the Fillmore had increased during the war, even as housing conditions deteriorated due to overcrowding and negligent landlords.[62] The Fillmore-based, Black-owned newspaper the *Sun-Reporter*, described poor conditions in the neighborhood in 1944, using the Double V campaign's critical framework: "Crowded 9, 10, 15 to a single room with only one window. They work all day for the maintenance of democracy and the defeat of fascism abroad and come home in the evening to face the realities of domestic fascist practices."[63]

Outside the extractive housing market in the Fillmore, Black newcomers to San Francisco—if they worked for the military—could find homes in new, hastily built military housing in Hunters Point. In response to a housing shortage during the war, San Francisco made use of federal funds from the Lanham Public War Housing Act to build fifty-five hundred "temporary dwelling units" (TDUs) in Hunters Point for shipyard workers, rapidly transforming former cattle yards and an undeveloped hillside into a residential neighborhood, with paved roads, sewers, schools, a health center, and a public library.[64] By 1945 war housing units in Hunters Point were home to approximately twenty thousand people, one-third of whom were Black shipyard workers and their families.[65] Working-class white ethnic residents of the Bayview, meanwhile, organized to preserve their neighborhood's whiteness; during the war, as Johnson reports, "handbills were distributed from door to door threatening to burn down the house of anyone selling property to Black people."[66]

At the end of World War II, then, most Black San Franciscans lived in the Fillmore and in temporary, military housing in Hunters Point. In the postwar decades—in the context of San Francisco's booming postwar economy—an expansive urban renewal project would lead to evictions from the Fillmore, while racist public housing practices continued to confine Black San Franciscans in Hunters Point. Meanwhile, in the 1950s and 1960s, in the context of demobilization at the Hunters Point shipyard and "white flight" into the growing Bay Area suburbs, some Black households were able to move into the Bayview, and the two neighborhoods were increasingly seen as one. In the mid 1960s, residents began referring to "Hunters Point-Bayview" and "Bayview-Hunters Point," reflecting a spatial expansion of a Black sense of community and place in southeast San Francisco.

At the same time, unbeknownst to most residents, a secretive laboratory had begun to operate on the Hunters Point Naval Shipyard: the Naval Radiological Defense Laboratory (NRDL). The NRDL grew out of U.S. experiments with nuclear testing in the Bikini Atoll in the Pacific Ocean in 1946, and the lab participated in subsequent nuclear tests while conducting radiological experiments on the shipyard until it closed down in 1969. Anti-Black housing discrimination confined Black San Franciscans in Bayview-Hunters Point at the same time that U.S. militarism entangled the neighborhood within the geographies of nuclear weapons production.

CROSSROADS IN SAN FRANCISCO

In 1945, a year after Johnson published *The Negro War Worker in San Francisco*, the United States dropped two atomic bombs on the Japanese cities of Hiroshima and Nagasaki. By the end of that year, the bombs had killed an estimated 210,000 people and left an untold number of burned and radiated victims. Almost immediately after the blasts, U.S. scientists traveled to the devastated cities to collect data on the effects of radiation from traumatized survivors. Simultaneously, the military prepared to detonate three more nuclear bombs at Bikini Atoll in the Marshall Islands, in the Pacific Ocean, as part of an experiment called Operation Crossroads. The Crossroads "test"—which for Bikinians who had lived on, and were removed by the U.S. military from, the atoll was a full-scale disaster—led to the establishment of the NRDL in Hunters Point. Through the NRDL, Hunters Point became a node in the geography of military nuclear research facilities and supported further atmospheric weapons "tests" in the Pacific Ocean and U.S. Southwest.[67]

Operation Crossroads sought to understand the material, biological, and environmental consequences of the atomic bomb for the U.S. military, and in particular the effects of the bomb on navy ships. In the summer of 1946, one hundred military vessels were placed in the middle of Bikini's lagoon, as the military "experiment's" main target. The first bomb, Shot Able, was dropped by aircraft and missed the target ships. The second bomb, Shot Baker, was detonated from ninety feet underwater, despite warnings from Los Alamos National Laboratory scientists that an "underwater test against naval vessels would create so many hazards that it should be ruled out at this time" and that "the water near a recent surface explosion will be a witch's brew."[68] In the official record and historiography of Operation Crossroads, participants expressed surprise at the aftermath of Shot Baker, when in fact the fallout was much as the scientists had predicted.[69] The bomb exploded from underneath the lagoon as a tall, wide column of water, followed by a circular wall of radioactive ocean mist that pushed outward, blanketing the lagoon with radiation. Radioactive fallout also contaminated hundreds of nontarget support ships outside of the lagoon and forced the military to formally end the Crossroads "experiment" while directing its efforts to containing the spreading radiation.[70]

Confronted with a fleet of radioactive warships in the Pacific Ocean, the U.S. military looked to Hunters Point, whose naval shipyard was already part of the geography of the bomb. Components of the atomic bomb dropped on the city of Hiroshima had been loaded onto a cruiser at Hunters Point in July 1945, and the shipyard had been the site of departure for many Bikini-bound ships the following summer.[71] In the aftermath of Operation Crossroads, Hunters Point became the center of the U.S. military's West Coast nuclear decontamination efforts. By December 1946, seventy-nine ships from Crossroads, including several of the target ships from the Bikini lagoon, had returned to the San Francisco Bay Area, primarily to Hunters Point. The attempt to decontaminate radioactive warships at Hunters Point was a practical necessity for the military, but it became another experiment, aimed at generating knowledge about radiological cleanup that would subsequently support future nuclear weapons tests. Radioactive ships from Bikini, docked at Hunters Point, were converted into floating labs to experiment with decontamination procedures; some of these ships were later used to store radioactive waste produced at laboratories and military bases across the Bay Area, and they were eventually sunk off the coast of California.[72] The military thus brought the radioactive disaster of Crossroads back to San Francisco.

Post-Crossroads experiments in nuclear disaster management on the Hunters Point Naval Shipyard were institutionalized with the establishment of the NRDL in November 1947. Archival materials from the lab's early years, in the late 1940s, detail its primary activities as involving contamination and decontamination studies; atomic bomb detonation observation and studies; monitoring procedures and standards; biological effects and clinical studies; and "personnel protection studies," which included establishing "tolerance levels" of radiation for "various tactical situations."[73] During the 1950s and 1960s, the NRDL further developed its biomedical studies through an extensive animal research program at the shipyard. Sheep, dogs, rats, mice, goats, donkeys, and frogs were bred, irradiated, and disposed of on the military base and in the ocean.[74] NRDL scientists also worked as radiological safety monitors during atmospheric weapons tests in the Marshall Islands and the Nevada Test site and brought back radioactive fallout to Hunters Point to use in laboratory experiments.

In its early years the NRDL grew quickly, employing 556 people by 1953.[75] Yet until a large, modern laboratory building was constructed in 1955, the NRDL conducted its radiological research in surplus military buildings across the shipyard, including a former barrack, where the animal colony was housed, and an old mess hall, where the physics department, instrumentation, and supply materials were crowded together.[76] The lack of space for the safe storage of radioactive fallout returning from nuclear detonations in Nevada and the Marshall Islands troubled NRDL scientists. According to a monthly progress report for the lab from 1947, "the problem of storage of samples, proper security of samples, and sufficient separation of active samples from sensitive research instruments is a very serious problem."[77] Finding adequate storage space for fallout remained a problem for years. During the late 1940s, fallout was kept in lead boxes in a small shack outside one of the laboratory's temporary buildings, which, as NRDL scientists themselves noted in a memo requesting better facilities, put laboratory technicians and shipyard workers (such as those contracted to dispose of the NRDL's waste) at risk.[78] Later, in their request for a "proper" isotope storage building in 1950, NRDL scientists explained that radiation from its storage shacks had risen to an unsafe level, exceeding permissible radiation exposure levels of the time.[79] These details indicate the haphazard nature of radiological experiments in the first eight years, at least, of the NRDL's operations at Hunters Point.[80]

The NRDL is only partially responsible for radioactive waste at the shipyard—radioactive materials were also commonly used in military shipbuilding, and thus by the Hunters Point Naval Shipyard's general operations, until the 1970s. Moreover, radioactive waste represents only a segment of the hazardous waste the military introduced into the soil and groundwater on the base, which includes lead, solvents, PCBs, asbestos, and other materials common to industrial shipbuilding at the time.[81] And more broadly, the shipyard is only one site within the geography of "industrial overburden" in Bayview-Hunters Point today.[82] Still, the NRDL's role in the military's nuclear research agenda locates Hunters Point within global geographies of U.S. military violence. Though technically devoted to nuclear defense, the NRDL actively contaminated the shipyard at the same time Hunters Point was becoming a majority Black neighborhood. Preparations for nuclear war thus became part of the subterranean ecology of the city and constitute the material basis for some of the more contentious environmental politics at the shipyard today.

RACIALIZED URBAN DEVELOPMENT
IN POSTWAR SAN FRANCISCO

With the postwar slowdown of operations at the shipyard and the expectation that wartime military dwellings would soon be demolished, war workers left Hunters Point en masse. Between 1950 and 1970, the total population of Bayview and Hunters Point declined by almost half—from 51,406 to 30,064 people. Yet it was primarily white households who were able to move. During those same decades, the white population in Bayview and Hunters Point declined by 59 percent, while the Black population grew by 17 percent.[83]

Several factors contributed to this differential mobility. First, white households had access to U.S. Federal Housing Administration mortgage loans and therefore to the postwar boom in suburban housing construction in the Bay Area, while Black households were systematically denied those loans.[84] Second, the San Francisco Housing Authority (SFHA), which took over the military's temporary dwellings in Hunters Point after the war, employed segregationist policies in the selection of tenants for its housing projects. In the 1940s and early 1950s, the SFHA contained Black tenants within the Westside Courts project in the Fillmore and Chinese tenants in the Ping Yuen project, in Chinatown. This racial separation allowed the SFHA to maintain all-white public housing projects in other parts of the city.[85] In defending its policy, the SFHA director, John Beard, insisted that public housing "must reflect the racial and ethnic composition of the neighborhood" (even though the "composition" of the city's neighborhoods was a product of historical racial and economic exclusions).[86] In 1953, a San Francisco Superior Court judge found the SFHA's tenant selection policies in violation of the equal protection clause of the Fourteenth Amendment.[87] By then, however, the agency had already reinforced historical patterns of public and private housing discrimination in the city. Moreover, the SFHA continued its segregationist policies even after the court ruling. When the SFHA built Alice Griffith Gardens in Hunters Point in 1962, for example, it rented 296 out of the development's 315 units to Black households.[88] Doing so would have provided homes for Black families looking to leave temporary military dwelling units in Hunters Point and for those uprooted by urban renewal in the Fillmore district. Yet it also would have allowed the agency to protect all-white public housing projects in other parts of the city. The SFHA thus actively shaped the racial division of space in postwar San Francisco.[89]

Postwar sectoral shifts in San Francisco's political economy reinforced the class dimensions of these racial urban geographies. Between 1958 and 1979, manufacturing employment in San Francisco declined by 25 percent (or roughly seventeen thousand jobs). At the same time, San Francisco experienced an "explosive growth" in finance, real estate, insurance (FIRE) and other service sectors. These sectoral shifts in labor and capital relied on and produced changes in the city's built environment, such as expanding the city's central business district to facilitate new flows of capital. In this same twenty-one-year period, more than fifty-two high-rise

buildings were built in the downtown area.[90] The SFRA's urban renewal
project in the Fillmore was part of this larger, postwar effort to restruc-
ture urban space and encourage the city's economic shift toward service
and finance capital.[91] The first phase of urban renewal in Fillmore began
in the late 1950s and took a massive eviction and clearance approach that
was likened by some Black Fillmore residents to a wartime bombing.[92]
The SFRA demolished entire city blocks, displaced thousands of people,
and forced many Black-owned businesses to close down.[93] Urban renewal
in the Fillmore did not go uncontested and by the mid-1960s Black residents and their allies had organized to change the trajectory of the urban
renewal project, including getting the SFRA to build more affordable
housing.[94] Still, much damage had already been done. Some displaced
Black Fillmore residents moved to Hunters Point, while others left the
city entirely.

In the post–World War II decades, then, the city of San Francisco
emerged as a corporate and financial services headquarters, and finance
and tourism became the city's two main industries.[95] At the same time,
the decline of manufacturing, wholesaling, and shipping contributed to a
growing landscape of abandoned land and informal waste sites in southeast San Francisco, leading to a new era of wastelanding. An economic
development report from 1975, for example, identified "the vast bulk of
the city's vacant industrial land . . . east of Third Street," near the Butchertown area of Hunters Point. Half of this land was used for open storage
and salvage yards. Environmental regulations, moreover—to the extent
they existed—went unenforced. According to the report, "Efforts . . . need
to be made to monitor and assure compliance with environmental standards by open-air uses and to minimize dirt and waste material generated on streets in the area from movement of materials to and from the
salvage, storage, and auto-wrecking yards."[96] By the 1970s, as skyscrapers
shot up downtown, wastelanding in southeast San Francisco assumed a
historically new form, as vacant industrial lots, informal waste practices,
along with increased industrial emissions. For example, in the early 1980s,
San Francisco expanded the Southeast Sewage Treatment Plant so that it
processed 80 percent of the city's wastewater while also contributing to
poor air quality in the neighborhood from the plant's diesel emissions (see
chapter 4).

Patricia remembered this landscape of waste when I interviewed her in 2011. I had asked about Heron's Head Park on the Hunters Point waterfront, near a recently shuttered power plant and a mile north from the shipyard. "Don't go there," she told me. "It's all landfill, it used to be where people dumped their old tires." In spite of new urban greening projects that aimed to transform parts of the industrial waterfront into spaces of recreation and nature enjoyment (Heron's Head Park, built in the 1990s, was the site of a failed freeway project from the 1970s; today it includes a reconstructed wetland habitat and is popular among Bay Area birders.) Patricia, who had lived and raised children in public housing near the power plant, the informal dump sites, and auto-wrecking yards along the waterfront, saw the new park as a place to stay away from. The park's trails and reconstructed marshes, and the assurances from state agencies that the park was safe to visit, could not erase her memories and lived experience. Her warning to stay away from the park indicates how toxicity also works through memory and subjectivity.[97]

Wastelanding is a relational process—environmental degradation and the marginalization of one place are linked with wealth accumulation and environmental health in another. As this chapter has shown, the social production of Bayview-Hunters Point as a Black, industrialized neighborhood was linked with the maintenance of white spaces elsewhere and with the city's postwar economic success. As geographer Jovan Scott Lewis writes in his ethnography of North Tulsa, "In the everyday context of Black spaces and their geographical iterations is the *making* marginal of Black place, with the protective order of White racialization calling that marginality into existence."[98] Lewis also observes that segregation is a geographic process of recognition, of "placing" people where they do and don't belong. In the second half of the twentieth century, hazardous waste facilities and Black people were seen to "belong" in Bayview-Hunters Point.

The differentiation production of urban space in San Francisco and the wastelanding of Bayview-Hunters Point took place through racial housing markets and public housing policies, war and migration, and shifting geographies of racial capitalism and settler colonialism. These processes produced a landscape in southeast San Francisco that subsequent environmental justice activism, in the 1990s, emerged from and sought to reshape. Yet wastelanding in southeast San Francisco also became the grounds for new forms of capital accumulation.

By the 2000s the Hunters Point waterfront had become the site of significant real estate interest, as companies and individuals aimed to profit from the transformation of contaminated industrial lots and the abandoned military shipyard into high-end, often "green" residential and commercial developments. The city's former wastelands had become a new frontier of accumulation.[99] As a 2006 *San Francisco Chronicle* article on Bayview-Hunters Point remarked, "When real estate developers look at the area's aging buildings, vacant industrial space and fallow rolling hills, they see dollar signs."[100] In the process, remediation and redevelopment companies accumulate value from a rent gap produced, in large part, by industrial toxicity, racial-spatial abandonment, and war. It is my contention that earlier colonial histories, beginning with Spanish missionization, are part of this history of devalorization and the making of what I consider a *toxic* rent gap—leading to gentrification predicated on a value differential produced by environmental degradation—as well.[101] This rent gap in Bayview-Hunters Point is doubly toxic because it was predicated not only on devalued and degraded land but also devalued and degraded lives.

Still, wastelanding in southeast San Francisco is not the only story to tell. In the following chapter, I describe Black counterplanning efforts in the 1960s and 1970s, which emerged in the context of widespread evictions from the Fillmore and the impending demolition of military housing in Hunters Point. Black counterplanning sought to build a livable neighborhood in Bayview-Hunters Point, by and for Black San Franciscans, during a period in which city agencies were actively shrinking Black urban space. Black counterplanning from the 1960s and 1970s also established a tradition of neighborhood activism that informs environmental justice organizing today.

2 Black Counterplanning for a New Hunters Point

The question of who is going to build what and for whom is the deep concern of the Bayview-Hunters Point Community Development Corporation.

—Osceola Washington, quoted in "Residents Start Housing Co-op," *Spokesman*, September 2, 1965

On the afternoon of Monday, September 27, 1966, a middle-aged, white San Francisco police officer killed a Black teenager, Matthew Johnson, in Hunters Point. The officer suspected Johnson and two friends of stealing a car. As Johnson ran away, the officer fired several shots at him, striking Johnson in the back. A nurse from the local Area Development office (a federally funded antipoverty organization) on Navy Road heard the gunfire.[1] She found Johnson near the bottom of a weedy hillside and attempted to save his life. A tense and growing crowd from the neighboring public housing project gathered near the site of Johnson's death as an ambulance drove his body away.[2]

By early evening, a crowd of young people—most of whom had, unlike their parents, grown up in public housing in San Francisco—assembled near the corner of Palou and Third Streets. They wanted answers. How would the police department handle the killing? Where was the officer who killed Johnson? Would he go to jail? Black men in Hunters Point were routinely arrested for petty crimes or simply for the offense of having an "attitude."[3] The crowd understood Johnson's murder as an effect of systemic anti-Blackness within the San Francisco Police Department, and they demanded that San Francisco mayor John Shelley come to

Hunters Point and address their concerns. Shelley initially refused. By the time Shelley finally arrived—late that evening, escorted by police in riot gear—he had pushed the crowd to their limit. Urban uprisings protesting anti-Black police violence had been erupting in cities across the country, including the Watts rebellion in Los Angeles the year before. The crowd at Third and Palou booed Shelley and threw eggs, old vegetables, and bricks. Later, writes historian Aliyah Dunn-Salahuddin, "small altercations took place between police and groups of youth. . . . [R]ocks and bricks were thrown at police and firemen and a car was set afire."[4]

The next day, in response, California governor Pat Brown sent two thousand National Guard troops to Hunters Point and to the Fillmore district, where the street protests had spread. In a striking image from the Black-run, Bayview-Hunters Point–based newspaper the *Spokesman*, troops armed with bayonet-tipped rifles march down Third Street.[5] At one point during the subsequent four days of demonstrations, police officers fired shots into the Bayview Community Center, where thirty unarmed youths had gathered.[6] Fortunately, in this instance, no one was physically harmed.

Many Hunters Point residents understood the youth-led uprising as decades in the making. As the editors of the *Spokesman* put it, "The death of young Matthew Johnson was not the basic reason for the rioting but a catalyst to what has been plaguing the Negro youth in poverty areas for years. No jobs, previous records of arrest, unfair hiring practices, inferior education and many others have been the underlying causes of the riot; probably before Matthew Johnson was born."[7] The death of a sixteen-year old represented a breaking point for Black San Franciscans struggling daily, for decades, with unsustainable living conditions.

A community memorial for Johnson, a few weeks after his death, also drew connections between police violence and other local living conditions—in this case, substandard public housing. Indeed, the San Francisco Police Department was rivaled only by the city's Housing Authority in the estimation of Hunters Point residents. On October 15, 1966, several hundred people participated in a "plant-in" (named in reference to the civil rights tactic, the sit-in) at the site of Johnson's killing. Volunteers spent the afternoon cleaning up the weedy ravine and planted dozens of small trees and shrubs. Reporting on the event, the *Spokesman* noted, "Many of

Figure 2. Photo of National Guard troops with bayonets fixed marching down Third Street in Hunters Point on September 28, 1966, in response to protests against the killing of Matthew Johnson. SOURCE: *Spokesman*, October 8, 1966.

the hundreds of Hunters Point residents who attended on Sunday, so deplored the desolation of the area that they felt something of this sort was needed."[8] The paper ran an image of a group of women gathered around the murder site, three of them bent over the ground, assembling an earthy memorial. While the plant-in memorialized Matthew Johnson and served as a form of community healing, it was also a grassroots beautification project that linked the police killing with the slow violence of unlivable, state-neglected public housing. The memorial rematerialized the site of Johnson's death from a vacant lot into a terrain of political struggle—foregrounding the underlying racism connecting both police killings and state absentee landlordism in Hunters Point, and altering existing racial-spatial arrangements through creative forms of placemaking.[9]

In this chapter I examine some of the ways Black Hunters Point residents worked to make space for Black life in San Francisco in the 1960s

Residents making memorial at murder site.

Figure 3. Memorial plant-in at the site of the murder of Matthew Johnson in Hunters Point, October 15, 1966. SOURCE: *Spokesman*, October 16, 1966.

and early 1970s. Confronted with untenable living conditions, Black residents pursued what I call *counterplanning* practices that aimed to produce what geographer Katherine McKittrick calls a "more livable, humane socio-spatial arrangement."[10] I conceive of "planning" broadly, in line with what geographer Ruth Wilson Gilmore calls the pursuit of "particular kinds of change in order to produce the conditions under which social and cultural reproduction might happen."[11] Planning practices aim to bring desired worlds into being. They are placemaking activities that target, or unfold across, a range of spatial scales. The memorial park for Matthew Johnson, however provisional, offered a different plan—a counterplan— for the abandoned lot and, by extension, for an abandoned community.

Black counterplanning in Hunters Point opposed the state violence of police killings, substandard housing, and the bulldozers and eviction notices that came with urban renewal. Counterplanners sought freedom from oppressive state practices by asserting community control over the conditions of social reproduction, including housing, infrastructure, food, and health care.[12] For example, a rent strike against the SFHA protested rats and cockroaches, indoor mold, and overflowing garbage cans, and envisioned a local housing cooperative instead, while the Hunters Point-Bayview Community Health Service critiqued discriminatory treatment at San Francisco General Hospital as part of the "dual system of public health" in the city and aimed to build an alternative health-care institution to serve neighborhood residents.[13] Black women were at the forefront of counterplanning for a more livable city in Bayview-Hunters Point. And they worked to repair and redevelop the built environment in ways that supported the reproductive labor of mothers and caregivers. Black counterplanning reflected and produced Black geographies, which, McKittrick writes, are "material and imaginative . . . critical of spatial inequalities, evidence of geopolitical struggles, and demonstrative of real and possible geographic alternatives."[14] Still, many counterplanning projects in Bayview-Hunters Point during this period were reliant on federal funding, largely through the War on Poverty. When these social and urban planning funding streams disappeared in the early 1970s, residents lost many of the resources that had supported their work of creating a more just city.

FREEDOM FROM THE HOUSING AUTHORITY

On the morning of March 8, 1966, Ollie Wallace, a tenant of the Alice Griffith Gardens public housing project in Hunters Point (located next to the Hunters Point Naval Shipyard) and a father of young children, heard a loud banging on his door. The SFHA had sent two movers and a sheriff to evict Wallace from his apartment. Ethelene Wilson, another tenant of Alice Griffith Gardens and a staff worker at the nearby Area Development office, noticed the "familiar van that is used in moving the furniture of evicted tenants."[15] Wilson promptly called the main Hunters Point Area Development office on Third Street, and the entire staff rushed to Alice Griffith Gardens, joining nearly one hundred people who had gathered to protest the eviction. The crowd sang "We Shall Overcome," the spiritual-turned-ballad of the civil rights movement, while Harold Brooks, director of the main Area Development office, negotiated with the Housing Authority. The previous year, John Beard—the SFHA director who had maintained an explicitly segregationist policy for public housing in San Francisco—was replaced by Eneas Kane. Kane promised many reforms, including an agreement to notify Brooks of any rent failures in Hunters Point before taking action to evict. In this instance, the SFHA Commission chairman, Steve Walters, acknowledged the agency had made a mistake and halted the forced eviction.[16]

For Hunters Point residents, the near eviction of Ollie Wallace and his family was part of a racist pattern of targeted abuse and neglect by the Housing Authority, not unrelated to the police violence that six months later killed Matthew Johnson. When not collecting rent or evicting tenants, the SFHA functioned as a typical absentee landlord, leaving public housing projects in Hunters Point in a state of disrepair that violated multiple municipal health codes. The *Spokesman* often commented on the visible neglect of Hunters Point public housing: "The trash is left in the street for days waiting for the long overdue street cleaning crew. The street cleaning crews are so seldom seen in our neighborhood that one gets the impression that they clean our streets in their spare time."[17] The editors of the paper were particularly incensed by the conditions of Alice Griffith Gardens. Constructed in 1962, the public housing development was named for a progressive white San Francisco social housing advocate, Alice Griffith, who had recently passed away.[18] Yet according to a 1965 article in the *Spokesman*, Alice Griffith Gardens was "one of the best examples of how short-sighted and unsympathetic the city can be in making accommodations for its less fortunate citizens."[19] The housing project was a "sardine can" that could lead, the paper warned, to a Watts-like uprising. Moreover, it was overrun by rodents. "Rats as big as coke bottles can be found running about . . . just before the sun goes down," one resident told the *Spokesman*. According to historian Keenaga-Yamhatta Taylor, in the 1960s "rats were the most visceral example of the unequal living conditions forced onto Black people" and "came to symbolize the degradation of Black urban life in the United States."[20] Across the country, Black residents living in substandard housing with poor sanitation services suffered rodent bites at such an alarming rate that President Lyndon Johnson sent the Rat Extermination and Control Bill to Congress in 1967. Congress rejected the bill.[21]

The Wallace incident motivated Hunters Point tenants to take further action against the Housing Authority, in protest of both the near-eviction and the agency's ongoing failure to provide decent housing. The following day, on March 9, 1966, sixty Hunters Point residents gathered to protest at an SFHA commissioner meeting. The protestors filled a small wood-paneled meeting room and spilled out into the hall, while a local news broadcasting station filmed the event.[22] During the demonstration, one Black man held up a sign reading, "We Don't Need Rats for Playmates" and placed himself strategically behind Commissioner Walters, in direct view of the television camera. At one point during the filming of the event, the man flipped his sign around, revealing the back side, which read, "Pied Piper, come for your rats." Walters ignored the tenants' demands to put Hunters Point public housing on that day's agenda and, after conducting the meeting, adjourned the session with a rap of his gavel. As Walters and other commissioners prepared to leave the room, however, calm voices announced, "Close the door. Lock your arms." Unmoving bodies of Black demonstrators blocked Walters from exiting and compelled him to sit back down in his chair, as George Earl, a public housing tenant and one of the leaders of the protest, read from a list of grievances. After Earl proclaimed, "We protest the disgraceful way that garbage men and Housing Authority employees leave trash around our homes," a Black woman from the crowd cried out, "They don't do it in Pacific Heights!" The woman's comment compared Hunters Point with a wealthy, white neighborhood on the other side of the city, thus critiquing the city's sanitation services and the Housing Authority as contributing to racially uneven development in San Francisco. After Earl finished his speech, the demonstrators sang "We Shall Overcome," as the crowd had done outside Ollie Wallace's house the day before.

The SFHA was part of the everyday life of most Hunters Point residents. At the time, the agency managed "the largest single concentrated block of public housing units in the city" in Hunters Point, which included the temporary dwelling units that had been built for shipyard workers during World War II.[23] And despite Kane's promised reforms, the SFHA's administration of public housing still required a constant struggle by tenants to monitor problems and demand basic living conditions, such as keeping their units up to housing codes the state itself had set. Residents understood the Housing Authority as part of a broader structure of anti-Black state violence in the city. As a local resident editorialized in the

Spokesman in 1967, "The difference between the Redevelopment Agency, the Housing Authority, and the police departments, is that one will kill you quickly, the other will leave you to the rats, and the other one, the Redevelopment Agency, will just leave you."[24]

A few days after the demonstration at the SFHA commissioners' meeting, tenants met at the Bayview Community Center, just off Third Street, to launch an organized effort to improve public housing conditions in Hunters Point. Their first order of business was forming a tenant union. Then they planned for a rent strike, which included setting up a trust fund to collect withheld rent. Tenants at the meeting also wondered whether they could use the trust fund to buy land on the hill from the SFHA and establish a local housing cooperative, which would provide autonomy from the SFHA and community control over the provision of housing in the neighborhood. And they voted on a new protest tactic to pressure the SFHA to take action in Hunters Point: jamming the agency's telephone lines by repeatedly calling, only to say, "We want to live in freedom" and then hanging up.[25]

On September 22, 1966, the newly formed Hunters Point Tenant Union (HPTU) put forward a list of grievances to the SFHA that included repairing holes in the exterior and interior portions of buildings; painting walls; cleaning up broken glass and debris; repairing broken windows; repairing defective lighting and plumbing; installing stairways so that tenants could access their apartments "without the necessity of climbing dirt barriers"; exterminating cockroaches and rats; and ending the practice of counting wages earned by school-aged children in determining monthly, subsidized rents.[26] Matthew Johnson was killed by the San Francisco Police Department five days later, and the ensuing street protests—in part, a response to unlivable housing—ricocheted through Hunters Point. In October, the Housing Authority applied to U.S. Housing and Urban Development (HUD) for emergency funds for repairs in Hunters Point, but the tenants' union wanted to pressure SFHA to act swiftly, and they went forward with the strike.[27] So, on November 1, tenants deposited their rent money into a trust fund, withholding it from the SFHA in protest. They were soon joined by tenants' unions across the city, including from Potrero Hill, North Beach-Chinatown, Valencia Gardens, Sunnydale, and the Western Addition.[28]

Figure 4. Political cartoon depicting the SFHA as a white landlord threatening to evict a young Black family. SOURCE: *Spokesman*, October 15, 1966.

On the eve of the strike, the *Spokesman* ran a political cartoon with the caption, "The Biggest Slumlord in the West." The hand-drawn image depicts the Housing Authority as an older, white landlord threatening a young Black family with eviction. The family stands defiantly in front of him, hands behind their backs. In the image, even the rats and cockroaches have joined the strike, in solidarity. In the background, windows of the barrack-style housing are broken or boarded up, while trash and broken bottles litter the street. The Health Department is closed. The cartoon represents the state as both extractive (in the caricature of an SFHA slumlord) and absent (in the closed doors of the Health Department).

Rather than waiting for the state to address the unsustainable conditions of public housing, the HPTU organized to repair public housing units on its own. Using some of the withheld rent, tenants set about making needed repairs, beginning with a fresh coat of paint on run-down

Figure 5. Sign announcing that the HPTU is repairing public housing units in Hunters Point, December 20, 1966. SOURCE: "George Earl on Hunters Point Tenants Union," aired December 20, 1966, on KRON-TV. Bay Area TV Archive, San Francisco State University.

buildings. In response, the SFHA filed a court injunction against the union, including charges against HPTU officials who, according to the SFHA, "were wrongly acting as SFHA agents by collecting and using SFHA rent money."[29] The SFHA won the injunction, and the HPTU ignored it. On December 20, 1966, a white television reporter interviewed George Earl in front of one of the housing projects. Earl described the capacities of residents to organize themselves. He showed the reporter how the HPTU had begun repainting the housing units, replacing "drab grey" with yellow—a brighter, more hopeful color. And he explained that the tenants' union was also contracting with an extermination company—getting rid of rats and other pests was a priority for the tenant-led repair effort. Handwritten signs hung on the side of one building, reading, "This building is being painted by the Hunters Point Tenants Union." Earl informed the journalist, "We've waited long enough. . . . We wanted to see what could be done, if tenants got together and wanted to see something done."[30]

The HPTU-led rent strike succeeded in its some of its immediate goals. The growing, city-wide momentum of the strike finally pushed the SFHA to locate funding for housing repairs in Hunters Point. HUD had denied the SFHA's earlier application for emergency funds, because the federal agency did not provide renovation grants for existing housing. In March 1967, five months into the rent strike, however, HUD approved SFHA's request to use city funds for repairs in Hunters Point. In April 1967, Mayor Shelley pledged $350,000, and the SFHA contributed an additional $150,000, to address some of the HPTU's demands (this amount was less than the $600,000 requested by the HPTU, to cover basic code violations). Further, the SFHA agreed to a separate demand, that Hunters Point residents be hired to work on those repair projects—thus addressing, if temporarily, high unemployment rates in the neighborhood. Following another court order, the HPTU transferred the escrowed rent money to the SFHA.[31]

The list of grievances compiled by the HPTU reveals the centrality of environmental health concerns to housing activism in Hunters Point. In the 1960s, in addition to rats, garbage emerged as a potent symbol of sociospatial marginalization in cities across the United States (indeed, uncollected garbage attracted rats). For example, organizations such as the Puerto Rican Young Lords, styled after the Black Panther Party for Self-Defense, politicized garbage in New York City's El Barrio/East Harlem neighborhood.[32] In the summer of 1969, the Young Lords launched a "garbage offensive" in El Barrio, addressing what residents saw as one of the biggest problems in their neighborhood: the absence of sanitation services, leading to overwhelmingly littered streets. The Young Lords initiated volunteer street-sweeping events on the weekends and soon adopted more confrontational tactics, such as placing piles of trash strategically in the middle of busy intersections, thus making the marginalization of El Barrio visible for commuters driving to and from Manhattan. Garbage had become a symbol of racially uneven development for El Barrio residents. Indeed, the political geography of garbage—where it accumulates, and where it does not—has motivated social movements around the world. Geographer Rosalind Fredericks shows how the "garbagescape" in Dakar, Senegal, became, through the youth-based Set/Setal movement, "a central terrain of contestation over the legitimacy of the Senegalese state."[33] As with the Young Lords and the Set/Setal movement, garbage in Hunters Point in the 1960s was both material and symbol—it smelled and attracted rats, and it represented an oppressive social system. In a full-page spread of its June 11, 1966 issue, the *Spokesman* celebrated the coming demolition of the much-despised war housing (discussed in the next section). Soon, the paper wrote, "some of the symbols of San Francisco's neglect for their forgotten citizens will be torn down." These symbols of neglect included piles of garbage. Captions accompanied the images on the page, such as "People have to live amid refuse (left by trashmen)" and "Garbage under their breakfast window." This last image of overflowing trash bins under breakfast windows provides a glimpse into the challenge of reproductive labor under conditions of racial capitalism and discloses the intimate and tactile nature of state neglect in everyday life.[34] It also indexes patterns of racially uneven development on the urban scale. Trash piled up in Hunters Point at the same time as the explosive growth of skyscrapers in downtown San Francisco and the boom in service and FIRE industries, discussed in chapter 1.

The HPTU rent strike and tenant-led housing repair project reflected the ways Bayview-Hunters Point residents, in Ruth Wilson Gilmore's words, "mingle[d] reformist and radical ideologies and strategies" in the social production of space.[35] On one level, the HPTU's strike was about holding the SFHA accountable and getting public housing in Hunters Point up to code. Yet the HPTU's vision was also ideologically expansive and included a critique of the SFHA as part of the racial state. The idea of forming a housing cooperative, the decision to use withheld rent to pay for infrastructural repairs, and the language of "living in freedom," moreover, reflected a desire for community control and Black self-determination.

While these desires are part of the history of Black freedom struggles in the United States, they took a particular shape and momentum within the historical context of the mid- to late 1960s, at the regional and national scales. In 1964, Bay Area civil rights organizations, led by young activists, began leading prominent, highly visible campaigns protesting discriminatory hiring practices in San Francisco, targeting hotels, auto dealerships, and the Bank of America's downtown headquarters.[36] These campaigns marked a shift toward civil disobedience tactics, such as sit-ins—and, in the case of the auto dealership, lying down under cars—which, as historian Robert Self writes, "emerged in the early 1960s as the hallmark of new left protest."[37] Organizers in the Fillmore also drew on these modes of collective action, such as lying down in front of bulldozers to protest urban renewal.[38] The occupation of the SFHA meeting in March 1966 and the HPTU strike between 1966 and 1967 adopted these direct action tactics.

On the national level, the year 1964 marked the beginning of President Johnson's War on Poverty—an ambitious yet deeply flawed legislative agenda, mostly carried out through the newly created Office of Economic Opportunity (OEO). Notably, Johnson's 1964 Economic Opportunity Act included the Community Action Program, which authorized local organizations and programs to be "developed, conducted, and administered with the maximum feasible participation of residents of the area and members of the groups served" (the four Area Development offices and many other social organizations in Hunters Point were funded by this program).[39] Scholars have shown how policy makers intended the Community Action Program paternalistically—to address the perceived incompetence and disorganization of poor urban neighborhoods—and how the concept of community action was closely linked with U.S. counterinsurgency efforts abroad.[40] The funding and political access offered by War on Poverty programs, moreover, tended to incorporate and institutionalize oppositional politics.[41] Yet in some places, as Self argues, the social infrastructure established by the community action mandate evolved "into a vehicle for community mobilization to challenge local political officials."[42] That is, as the failures of the federal War on Poverty became apparent to many urban communities in the second half of the 1960s, some local organizers transformed federal antipoverty agencies into autonomous, or at least locally controlled, institutions, which appropriated some of the language of community action—aligning it with notions of community control and self-determination—and confronted municipal power structures. Self describes two variants of self-determination that emerged in second half of the 1960s: the revolutionary nationalism of the Black Panther Party for

Self-Defense and War on Poverty grassroots community empowerment. Founded by Bobby Seale and Huey Newton in 1966 in Oakland, the Black Panthers had a presence in Hunters Point; for example, they ran a free breakfast program, delivered speeches, and accompanied the coffins of Black men killed by police officers.[43] Yet the self-determination articulated by Hunters Point residents themselves emerged primarily from organizers working through antipoverty and other local service programs. For example, George Earl, one of the leaders of the HPTU rent strike, was also the head of the Hunters Point Inter Block Council, a neighborhood-level organization, established in July 1964, that coordinated the activities of nine block clubs in Hunters Point. The block clubs organized cleanup campaigns and youth basketball programs, and sought improvements from city agencies such as the SFHA and school district; more broadly, explained the *Spokesman*, block clubs aimed to "generate a feeling of pride and dignity among residents of the area."[44] Earl also served as vice chair of the Bayview-Hunters Point Community Development Corporation and was on the board of the *Spokesman* (its tagline was "Serving the People of Hunters Point-Bayview").[45] Both organizations were federally funded, through the OEO. The newspaper encouraged residents to participate in antipoverty organizations and programs in Hunters Point, as vehicles for community power, even as its editors maintained sharp critiques of city agencies, such as the Housing Authority, and of the federal War on Poverty itself. Self's description of antipoverty politics in Oakland is thus apt for part of the Hunters Point political landscape as well: "The community action mandate of federal antipoverty legislation . . . helped establish a new constituency for activists who attempted to mobilize the ghetto with a populist version of self-determination. Less Marxist-Leninist than urban grassroots insurgency, this variation on black power stressed the return of power to local communities (the 'people' or 'the community') from the hands of an assortment of outside institutions that included private capital, business, [and] the state."[46]

So, when the SFRA announced an urban renewal project in Hunters Point, residents responded by drawing from these emerging discourses and funding streams, with the goal of asserting control over redevelopment in their neighborhood. Led by Black women, residents envisioned an urban renewal project that would create space for, rather than displace,

Black San Franciscans. Specifically, they insisted that nothing would happen in Hunters Point without their consent. If their plans for urban development in Hunters Point never fully materialized, it was not, I argue, for lack of effort or vision, but because what they desired—to replace dilapidated war housing with a modern, permanent neighborhood—ultimately placed them in a dependent relationship with city agencies and vulnerable to broader shifts in national politics.

BUILDING A NEW HUNTERS POINT

The SFRA turned its attention to Hunters Point in the early 1960s, compelled by the California state legislature to demolish all wartime temporary dwelling units by 1970. The Ridgepoint and Wisconsin war housing projects on Hunters Point hill, totaling 2,727 units of housing, were designated "Redevelopment Project Area G" in 1963. The SFRA planned to use federal funds, available to cities through the 1949 Housing Act, to demolish the military dwellings and encourage private developers to build new homes. Neither the SFRA nor city officials appeared interested in protecting Black residents from displacement through urban renewal at first (indeed, in 1963 the Redevelopment Agency was well into its project of demolishing and clearing large swaths of the Fillmore). In a letter penned two years earlier, in 1961, for example, M. Justin Herman, the director of the SFRA, envisioned 117 acres on Hunters Point hill renamed "Ridge Point" and sold to families of "moderate income." He hoped the agency's urban renewal project would "assure an attractive urban design in an area with a sunny climate and outstanding view potentialities" and compared his vision for Hunters Point to another urban renewal project in San Francisco, in Diamond Heights, located uphill from the Glen Canyon and Noe Valley neighborhoods.[47] When it was completed, 20 percent of housing on Diamond Heights was federally subsidized, yet, as urban planning scholar Susan Fainstein writes, "Many of the market-rate houses and condominiums constructed sold at the upper end of the luxury housing market."[48] Herman believed urban renewal in Hunters Point would "add substantially to the property tax base of the city," yet this would have meant prioritizing housing options unavailable to most Black households who lived in Hunters Point. A 1965 San Francisco Board of Supervisors resolution underscored the underlying anti-Blackness of the urban renewal project when it emphasized "the desirability of replacing most of the substandard war housing with modern and moderately priced developments which will attract additional white people to Hunters Point and abolish the ghetto atmosphere."[49] In response, Bayview-Hunters Point residents organized to control the redevelopment process and ensure that affordable housing units—built by and for existing Black residents—were constructed instead, along with new infrastructure and an industrial park replacing the old Butchertown district, thus creating local jobs. To be clear, the temporary, run-down war housing was despised by current residents. As Texas-born Hunters Point resident and organizer Elouise Westbrook told a reporter in 1968, "Unlike other redevelopment areas in the city, [Hunters Point] is unanimously in favor of massive destruction, and the construction of new houses, schools, parks and places of employment."[50] With the scheduled demolition of thousands of housing units, however, the question loomed: Who would be able live in Hunters Point after redevelopment was complete?[51] Most Hunters Point residents seemed to want to stay in the neighborhood. In 1966 for example, a local block club conducted a door-to-door survey of three hundred households in Hunters Point. As the *Spokesman* reported, "The [survey] response not only revealed a desire to continue living in the area, but showed that residents also felt that they should have a voice in what is to become of the land, and what kind of housing will be built there."[52]

So Bayview-Hunters Point residents formed their own housing and redevelopment organizations, with the goal of establishing community control over urban renewal. In September 1965 (a year before the HPTU rent strike), Black residents met at Ridgepoint Church on Hilltop Road and established the Bayview-Hunters Point Community Development Corporation (CDC) to provide an institutional structure to control, rather than be subjected to, large-scale redevelopment. The CDC board (membership was open to all Bayview-Hunters Point residents) quickly got to work, hosting workshops on possible legal avenues for the new organization and conducting a community needs survey. Osceola Washington, who had moved to San Francisco from Little Rock, Arkansas, in 1944, served as the Bayview-Hunters Point CDC's first chairperson.[53] According to the *Spokesman*, "Under the chairmanship of Mrs. Osceola Washington, residents were active in the Bayview-Hunters Point CDC efforts [sic] to see that outside speculators or Urban Renewal groups are not allowed to buy or lease the land without regard for the thousands of people who now live there." As Washington put it, "The question of who is going to build what and for whom is the deep concern of the Bayview-Hunters Point Community Development Corporation."[54] The primary goal of the Bayview-Hunters Point CDC—similar to the idea of a housing cooperative, debated by the HPTU the following year—was to mobilize residents and to "buy land to build homes for the people of Hunters Point."[55] As the *Spokesman*, in an article on the CDC, explained to its readers, land in Bayview-Hunters Point was quite valuable. "Private and public interests from outside the area are already active in trying to obtain the land for their own interests. The people of the Community, however, do not need to leave it up to others. The kind of housing, commercial facilities, and recreation centers which will rise on the hill can be determined by the citizens of Bayview-Hunters Point."[56]

Leadership on the Bayview-Hunters Point CDC overlapped with another powerful redevelopment organization led by residents, the Joint Housing Committee (JHC). The Hunters Point Area Planning Board designated the JHC as "the official body to deal with the Redevelopment Agency in the planning and execution of any project" in 1966.[57] Whereas the Bayview-Hunters Point CDC aimed to own and manage housing and was autonomous from city agencies, the JHC had the primary function of working with the SFRA on the overall concept for urban renewal in Hunters Point. The JHC was instrumental in transforming the SFRA's original project into a different plan for Hunters Point, one that prioritized more affordable housing, job creation for local residents, and the needs of caregivers.[58] As the JHC constitution—written in 1966, in the context of anti-redevelopment organizing in the Fillmore—insisted, "All people currently residing in the area will have an opportunity to remain if they choose to do so. The new housing will be built so they can afford to live in it."[59] The JHC also pushed the SFRA to contract Black architects to design the new housing—ultimately three of the five developments were designed by Black architects—and planned to hire local Black workers to build them.

The new urban renewal plan for Hunters Point, as a joint project between the JHC and the SFRA, outlined redevelopment in two phases, constructing 2,200 housing units overall. The first phase envisioned 672 housing units built on vacant land on Hunters Point hill. These units would then serve as relocation housing for Hunters Point residents leaving soon-to-be-demolished war worker housing. The urban renewal plan also included the construction of two new schools, four childcare centers, a new shopping center, two new churches, community recreation and meeting centers, the undergrounding of utilities, and tree-lined walkways.[60] Laundry facilities would be located near playgrounds, so that mothers and other caregivers could watch children while attending to other reproductive labor. Moreover, each of the five housing projects in the first phase of redevelopment was to be managed by a local nonprofit sponsor, rather than by an outside company or the SFHA. Ridgepoint Church, Bayview-Hunters Point Credit Union, Bayview-Hunters Point CDC, Double Rock Baptist Church, and All Hallows Church—all locally owned or operated nonprofits—would replace the Housing Authority as managers of social housing in Hunters Point, providing a form of community control over a basic site of social reproduction, the home.[61]

The second phase of redevelopment included plans for more housing along with a light industrial park to replace the aging Butchertown district, with the goal of attracting labor-intensive industries that would provide jobs for Hunters Point residents. The JHC and SFRA also anticipated the creation of additional jobs for local residents by the navy, which at the time had plans to upgrade its shipyard and concentrate West Coast military shipbuilding and repair operations in Hunters Point.[62]

State agencies were fickle partners, however, and the JHC-SFRA Hunters Point urban renewal project experienced many delays. For one thing, the SFHA initially refused to release the military's former dwelling units to the SFRA, which the Housing Authority had purchased in 1954, without being paid $600,000 for the land. The SFHA's insistence on payment for 104 acres on Hunters Point hill threatened the entire redevelopment project, since federal funding was conditional on the city's pledge that it would make the land available at no charge.[63] Elouise Westbrook eloquently responded to the SFHA's intransigence on this issue: "It will, of course, be necessary for the Housing Authority to dispense with the old bugaboo of property rights and let human rights and needs abide."[64] Fortunately for Hunters Point residents, the San Francisco mayor's office disputed the SFHA's claim, and the land was eventually handed over to the Redevelopment Agency. More significantly, HUD rejected several of the city's applications for urban renewal funds for Hunters Point, and the SFRA—at least from the perspective of Hunters Point residents—dragged its feet in returning those applications to HUD and advocating for redevelopment in Hunters Point.

The women running the Bayview-Hunters Point JHC fought these obstacles each step of the way. For example, in April 1968 HUD rejected an application for a $60 million grant to support urban renewal in Hunters Point, citing the city's failure to secure its share of funding for the project.[65] In response, Lillian Woods, then chairperson of the JHC, met with Mayor Joseph Alioto on September 5, 1968, and presented him with a list of demands. In a letter outlining these demands, which was also addressed to Herman and to HUD regional administrator Robert Pitts, Woods insisted that the mayor's office find the money to support the city's share of project costs and return the urban renewal application to HUD, adding that "a sales tax is *not* an acceptable solution as it places a heavier burden upon poor people." "HUD and the City must keep their commitment to the community," Woods wrote, not only to build housing and create jobs in Hunters Point but also to build "new schools, a shopping center, parks, community facilities, pedestrian pathways, tot-lots, churches, nursery schools, etc., as recommended by the Joint Housing Committee in its planning the past two years." At the end of her letter, Woods asserted: "HUD and the City, if unable to fulfill Items 1-6, must reimburse members of the community their taxes paid so diligently all these years, and the City must repay community people for all their time and expenses wasted in planning for the New Hunters Point Community."[66] Woods's letter indicates how the JHC operated in both collaborative and oppositional modes in relation to the SFRA. The JHC's relationship with the SFRA represented access to needed federal resources to remake their neighborhood, but the agency was never to be entirely trusted. In this case, Woods's letter seems to have been effective; within a month of meeting with Woods, Mayor Alioto and regional Administrator Pitts reached an arrangement to secure the federal grant so that the Hunters Point urban renewal project could move forward.[67]

JHC
Executive Committee

Mr. Reuel Brady
Chairman, Labor & Industry

Mrs. Julia Commer Chairman,
Priority Certificate & Relocation

Mrs. Beatrice Dunbar
Chairman, Education

Mrs. Bertha Freeman
Chairman, Personnel

Mrs. Lessie Hopkins
Chairman, Community Facilities

Mrs. Espanola Jackson Rich
Chairman, Welfare Rights

Mrs. Mary Rivers
Parliamentarian

Mrs. Evelyn Snelgro
Chairman, Contract Review

Mrs. Elouise Westbrook
Chairman, Joint Housing Committee

Mrs. Marcelee Cashmere
Vice chairman, Joint Housing Committee

Mrs. Essie Webb
Chairman, Rehabilitation

Mr. George Williams
Chairman, Audit

Mrs. Ruth Williams Chairman,
India Basin Industrial Park

Figure 6. Members of the JHC Executive Committee in 1969. SOURCE: Bayview-
Hunters Point Joint Housing Committee and San Francisco Redevelopment Agency,
Hunters Point and India Basin Industrial Park calendar, Joseph L. Alioto Papers,
1958–1977, box 17, San Francisco Public Library, San Francisco.

The JHC and SFRA hosted a groundbreaking ceremony for the five new
housing developments on Hunters Point hill on Saturday, November 1, 1969.
Dr. Arthur Coleman—a prominent and respected Black physician, who
was also part of the city-wide Economic Opportunity Council leadership—
officiated. A calendar spanning October 1969 through September 1971—
the expected time frame for the first phase of redevelopment—was passed
out at the event.[68] Thick with text and images, the calendar included landscape designs and street layouts for each of the five new housing developments. The calendar booklet also included photographs of various JHC
committees and described their work over the previous three years. On the
calendar itself, nearly every day of the week was marked by an important
moment in U.S. Black history, including Black Californian history. This
included the 1855 San Francisco–based movement for the right to testify
in court (which had been denied to Black Californians by the original state
constitution) and the birth of Mary Ellen Pleasant, a Black abolitionist
who harbored fugitive enslaved people in her San Francisco home in the
1850s and helped fund John Brown's raid on Harper's Ferry in Virginia in

1859.[69] In doing so, the calendar interpreted the repair and redevelopment of Hunters Point as part of the history of Black political activism in the United States.

The JHC worked directly with the SFRA, as an official advisory committee. In contrast to other organizations, which saw themselves as autonomous from the state, or at least from specific state agencies—ranging from the Bayview-Hunters Point CDC to the Black Panther Party—the JHC was not separate from the state; indeed, it was, in a sense, part of the state. However, as is clear from the rhetoric of Eloise Westbrook and Lillian Woods, the women who led the JHC were not co-opted. Rather, they maintained a strategic relationship with the SFRA, taking advantage of the community participation mandate that accompanied federal support for urban renewal in Hunters Point and perhaps the SFRA's desire to remake its image in the context of anti-redevelopment organizing in the Fillmore. Organizers such as Westbrook and Woods recognized that they needed the power of the state to accomplish large, capital-intensive projects—namely, the demolition of thousands of temporary war dwellings and the construction of an entirely new neighborhood. By working from within the apparatus of municipal government while maintaining a critical perspective on this relationship, the women leading the JHC sought to leverage state power to reshape the built environment in Bayview-Hunters Point.

COMMUNITY CONTROL OVER FOOD AND HEALTH CARE

In 1960, to protest the refusal by businesses on Third Street to hire Black workers, Hunters Point residents formed the Hunters Point-Bayview Citizens Committee. The Citizens Committee organized events such as pickets at offending stores and demanded fair hiring practices for Black residents and higher-quality merchandise for local shoppers. The *Spokesman* later reflected on the campaign: "It wasn't long before the picketers were given minimal cooperation, that is, a token number of Negroes were hired in a couple of businesses. The group was then confronted with the question, how can we do for ourselves what others will not do?"[70]

The Neighborhood Co-op grocery store emerged in response to this question. Many of the businesses that had refused to hire Black workers were small grocery stores, and the Citizens Committee began to reimagine how Black Bayview-Hunters Point residents could buy food in ways that supported the neighborhood and the Black people who lived in it. In 1961 they began planning for a local food cooperative. The co-op started selling shares in 1963, and the grocery store opened two years later, in 1965. The following year, two thousand households owned shares in the co-op. In addition to providing quality, reasonably priced food, the co-op hosted an arts festival, maintained a book section, and served free coffee from Thursday to Saturday. The co-op thus offered an alternative to a food system that demeaned Black San Franciscans and represents an example of what geographer Ashanté Reese calls "geographies of self-reliance that unfold within spatialized food inequities."[71]

Essie Webb, who had moved to Hunters Point from Arkansas in 1944 with her husband, a shipyard worker, told the *Spokesman* in 1966: "One of the greatest achievements of which I was a part and now a member of is our Co-op Supermarket. This is a very good example of what we as a group of people can do."[72] Webb was active in many Bayview-Hunters Point neighborhood organizations; she served on the board of the Bayview Community Center and the *Spokesman*; was on the advisory committee of the Youth Opportunities Center and later on the executive committee of the JHC; and served as secretary of the Bayview-Hunters Point CDC and organized with her local block club. Given all these involvements, it is notable that she valued her role in establishing the food co-op so highly.

Black self-determination through building local, alternative institutions was also a central goal of the Hunters Point-Bayview Community Health Service (CHS). Seeking health care could be a dehumanizing experience for Black Bayview-Hunters Point residents. The nearest hospital was San Francisco General (located north of the neighborhood, in Potrero Hill), but many residents avoided it. As a survey conducted in the 1960s found, "41% of people [in Bayview-Hunters Point] would not go to SF General even if they were sick: because of the way they were treated, the long waiting lines, the inconsideration for them as individuals with dignity."[73] The CHS, which was founded by Dr. Arthur Coleman in 1967 (Coleman had practiced in Bayview-Hunters Point since 1948, although he did not live there) and funded by the U.S. Public Health

Service, offered a response to these demeaning experiences. Coleman conceptualized the CHS together with a group of Black residents from the neighborhood, and residents later provided the staffing and outreach teams for the health service.[74] In establishing the CHS, Coleman asserted that "health is a right, not a privilege" and that "there should be one level of health care for all of our citizens."[75] Moreover, as he emphasized in the first issue of the *Hunters Point-Bayview Community Health Service Newsletter*, "Black people must control their own destiny, especially in matters of health."[76] To that end, the CHS offered an alternative to SF General by providing free medical, dental, and optometry care to residents of southeast San Francisco. Coleman and the CHS staff recognized numerous racial, gendered, and economic barriers to accessing health care, and so the CHS also offered services such as free babysitting and transportation for parents (usually women) going to doctor's appointments, pharmacies, or visits to state welfare agencies.[77] And CHS staff members accompanied Bayview-Hunters Point residents to their appointments at other hospitals and city agency offices, as patient and client advocates.[78]

In her study of the Black Panther Party's health politics, sociologist Alondra Nelson argues that health should be "understood as an important feature of a broader conceptualization of the civil rights movement."[79] The Black Panthers, for example, aimed to "serve the people body and soul," in part through free medical clinics and free breakfast programs for children, including one located at Ridgepoint Church in Hunters Point.[80] Similar to, though not the same as, the Black Panther Party's health programs, the Hunters Point-Bayview CHS focused on a model of health care that addressed Black residents' social environment, rather than seeing only individual pathologies, and recognized that health rights for Black San Franciscans could only be achieved by addressing underlying social structures.[81] This change would not come from within the current healthcare system in San Francisco, but by building Black-led alternatives to it.

A central goal of this chapter has been to emphasize historical, Black-led visions of urban repair and redevelopment in Bayview-Hunters Point. I call these projects, broadly, Black counterplanning, because they opposed exclusionary urban policies and practices and aimed to bring other, desired worlds into being. Black counterplanners in Bayview-Hunters Point sought freedom from oppressive state practices through community control over the conditions of social reproduction, such as housing, infrastructure, food, and health care. Clyde Woods writes that "planning has always held the promise of creating new social relations based on economic redistribution, environmental sustainability, and the full realization of basic human and cultural rights."[82] Woods reminds us, however, that planning tends to reproduce social inequalities and oppressions when not informed by alternative conceptions of development based in Indigenous, working-class, and racially marginalized social groups. In *Development Arrested*, Woods finds the origins of an alternative, more just form of development in the Mississippi Delta region, "among the scattered, misplaced and often forgotten movements, projects, and agendas of its African American communities and of other marginalized groups."[83] The visions of the HPTU, the JHC, the Bayview-Hunters Point CDC, the Neighborhood Co-op, the Hunters Point-Bayview CHS, and other local organizations, I argue, similarly represent an agenda for a just urban development, by providing alternatives to the racist policies and practices of state agencies.

As historian Rhonda Williams argues, Black women have been at the center of national efforts demanding "the right to adequate housing, income, medical care, food and clothing," and their efforts "represented another phase of the black freedom struggle."[84] This characterizes the role of Bayview-Hunters Point women during the 1960s and 1970s as well. Elouise Westbrook, for example, served as chair of the JHC and was also community relations liaison for the CHS. Bertha Freeman served on the JHC and worked as a consumer advocate with the Bay Area Neighborhood Development organization (advocating for consumer protection), which had its offices in the Neighborhood Co-op. Marcelee Cashmere coordinated the Mothers Helpers unit of the CHS and served on the board of directors for both the Neighborhood Co-op and the *Spokesman*. Julia Commer was one of the cofounders of the Neighborhood Co-op and also served on the JHC, as president of the Burnett School Parent-Teacher Association, and on the advisory board of the Youth Opportunity Center. Commer's involvements in particular reflect how mothering in Hunters Point involved institutional leadership along with the everyday labor of social reproduction.

These names represent only a few of the women in local leadership positions during this time, but their varied involvements provides a glimpse into the ways all of these issues were seen as interconnected.[85]

CUTBACKS

The vision of a "new Hunters Point," as depicted in the 1969 JHC-SFRA calendar, was never fully realized. In May 1970, HUD funding delays prompted the SFRA director, Justin Herman, to fly with Elouise Westbrook and a dozen other members of the JHC to Washington, D.C., with the goal of persuading HUD to release $40 million that had been promised to Hunters Point in 1967. All accounts of this trip center on the negotiating skills of Westbrook in brokering a deal.[86] According to the *Sun-Reporter*, "Mrs. Westbrook and members of the Hunters Point delegation, Geneva Wheatfield and Mrs. David [Julia] Commer, led an exhaustive debate which sent Washington officials into backroom caucuses. Two hours later the verdict was reached which assured the release of federal funds to the workable redevelopment program."[87] Westbrook and others on the JHC won this particular battle, and the redevelopment project seemed poised to move forward after yet another hurdle. On October 25, 1972, a local television reporter interviewed Westbrook in front of one of the new housing sites. Against a background of wood frames and the sound of hammering, Westbrook spoke about the importance of a new light industrial park at Butchertown—part of the second phase of redevelopment for Hunters Point. The industrial park would provide permanent, rather than temporary, jobs for Black residents, she told the reporter, and allow people to "live here for ever and ever."[88]

A week later, Richard Nixon won reelection to the U.S. presidency in a landslide victory. Nixon's second administration took his win as a mandate to dismantle the programs created under the aegis of the War on Poverty. During the 1970s, moreover, the conservative critique of state welfare—that it was an entitlement or privilege, rather than a means to ensure basic human rights—became bipartisan consensus.[89] Broad cuts to Johnson-era welfare and urban renewal programs were announced in 1973. The effect of these cutbacks on urban planning and other social programs in Bayview-Hunters Point was severe. A major funding source for redevelopment in Hunters Point came from the Model Cities Demonstration Project, a program within Johnson's War on Poverty legislative agenda. In 1974 the Model Cities budget for San Francisco decreased from $7.35 million to $1.14 million, with the expectation that the remaining funds would close out the program entirely.[90] With the loss of this essential funding source, the second phase of redevelopment in Hunters Point was suspended, again.[91]

The *Sun-Reporter* contextualized this economic blow within a new policy landscape that undermined many 1960s-era social projects, while channeling even more money to the U.S. military. "Located in a section of the City which decision makers seldom visit and usually never live in, Hunters Point has borne the burden of Nixon's cutbacks in domestic spending. President Nixon has spent nearly five years dismantling viable domestic programs in the realm of education, medicine, and housing while giving the military a blank check."[92] Notably, although Nixon increased military spending, his administration's budget effectively abandoned many Hunters Point residents to the poorly built World War II war worker housing units on Hunters Point hill—the detritus of an earlier war. A few years earlier, the *Spokesman* had drawn connections between domestic housing policy and military geopolitics with an image demanding the state "Build Better Homes, Not Better Bombs" (prioritizing military expenditures was by no means unique to Nixon's administration). Published in 1967, the same year as the HPTU rent strike, the image links the unlivable state of housing in Hunters Point with the U.S. war in Vietnam. The demand to "Build Better Homes, Not Better Bombs" also recalls the Double V campaign from the 1940s, which connected fascism abroad and racism at home. The graphic and its commentary indicate another way the struggle for decent housing in Bayview-Hunters Point was understood within a critique of the racial state, in this case, U.S. imperialism.

Then, on April 17, 1973—a few months after President Nixon announced large cutbacks to federal programs funding jobs and social programs in Hunters Point—the secretary of the navy announced the Hunters Point Naval Shipyard would close down. Shutting down Hunters Point was part of a larger action to consolidate, reduce, and close military bases in thirty-two states. Yet Hunters Point, with its significant

Figure 7. Graphic demanding that the state "Build Better Homes, Not Better Bombs." SOURCE: *Spokesman,* October 1967.

job losses, was one of the more impacted locations.[93] The shipyard was a major employer of Hunters Point residents (and the largest industrial employer in the city), and its closure left hundreds of people from the neighborhood unemployed. The shutdown also impacted the local neighborhood economy, as businesses struggled in the absence of thousands of shipyard workers who had patronized shops and restaurants in the neighborhood.[94] The shipyard's closure thus compounded the harsh impacts of other federal cutbacks at the time, in social welfare and urban redevelopment spending.

The announcement of the base closure surprised and dismayed both local residents and city officials. The city had counted on military jobs to provide an economic base for the struggling neighborhood and to support the broader goals of urban renewal. As a news broadcaster reported, "What the city of San Francisco is particularly concerned about is that all their efforts to rehabilitate housing in the Hunters Point area will be for naught unless they [residents] can have jobs, and the logical place for jobs is here at the Hunters Point Naval Shipyard."[95] Meanwhile, the devolution of social welfare to cash-strapped cities was already rendering hollow a decade of promises made to Bayview-Hunters Point by San Francisco city agencies as well as by the federal government. In an application to the U.S. Economic Development Administration, for example, appealing for federal help to reuse the shipyard and provide for local, blue-collar employment, Mayor Joseph Alioto wrote that the "announced closing of the HPNS [Hunters Point Naval Shipyard] added to the bitterness engendered by the series of what community residents consider the heartless abandonment of their neighborhood by the federal government."[96]

Yet Alioto's strongly worded statement failed to mention the ways Bayview-Hunters Point residents already felt abandoned by San Francisco city agencies, such as the Housing Authority, Redevelopment Agency, and sanitation services—an abandonment residents had fought via assiduous counterplanning efforts throughout the preceding decade. Alioto's statement also deflected municipal responsibility for the effects of postwar economic restructuring in San Francisco around finance and tourism, an economic trajectory that left many Hunters Point households reliant on the very jobs—from the military, the War on Poverty, and the manufacturing industry—that declined or disappeared in the 1970s.

Ultimately, the closure of the Hunters Point Naval Shipyard has posed questions about the future of Bayview-Hunters Point and the place of Black residents in determining that future. In the following chapter I show how the neighborhood tradition of organizing for community control has faced new challenges from the military and the federal Superfund program.

3 The Politics of Environmental Repair

Though the U.S. Navy shut down its operations on the Hunters Point Shipyard in 1974, it continued to own the military base, and during the 1970s and 1980s it leased out parcels to private businesses and government agencies. The navy's main tenant during this time was Triple A Machine Shop, Inc., a private ship repair firm. In 1985, in the context of a disagreement over Triple A's lease, the navy discovered that the firm had been dumping hazardous waste on the military base. Subsequent investigations led the San Francisco district attorney to file sixteen felony charges against the firm for burying PCBs, asbestos, solvents, heavy metals, and other toxic materials on the shipyard, in violation of California's Hazardous Waste Control Act of 1972. The company was found guilty on five criminal counts.[1] Following this lawsuit, in 1989 the EPA classified the Hunters Point Shipyard as a Superfund site—a regulatory category designating highly contaminated properties that require environmental remediation. The district attorney's lawsuit and the shipyard's new regulatory status contributed to the navy's decision, in 1991, to officially close the shipyard and begin the process of transferring the land back to the city of San Francisco.[2]

As Bayview-Hunters Point residents would soon discover, Triple A was by no means the main source of contamination on the military base. For

decades the navy had routinely dumped its waste products in San Francisco Bay and buried them on the grounds of the shipyard. When the navy
decided to hand over the shipyard to the city for civilian reuse, it sought
to transfer land saturated with some of the most hazardous substances
produced in the twentieth century, including radioactive waste from the
NRDL.[3] The NRDL had shut down in 1969, the same year Eloise Westbrook, Lillian Woods and others on the JHC celebrated their hard work
planning for a "new Hunters Point" at the groundbreaking ceremony on
Hunters Point hill. Although the radiological laboratory's staff left the military base that year, the lab's waste products remained. Today, these two
histories—of war's ecologies and Black counterplanning—collide in the
politics of base cleanup and redevelopment.

The navy's decision to release the shipyard to San Francisco led to urgent questions about how to incorporate a highly toxic military base into
the city's spatial and political economy, and who would have the power
to make decisions about this process. Was it even possible to clean up the
shipyard? How would the city reuse the land? How much control should
residents of Bayview-Hunters Point—with the neighborhood's history of
racist housing policies, failed development projects, and toxic exposure—
have over this process? In 1999, following the SFRA's adoption of a reuse
plan focused on housing and office space, Mayor Willie Brown granted the
development rights for the shipyard to the Florida-based home-building
company Lennar Inc.[4] Under the terms of the deal, Lennar would pay
nothing for the land but would make upfront investments in the infrastructure (such as grading roads and installing new street lights) needed
for the new urban development.[5]

The SFRA redevelopment plan offered a different development path
for the historically wastelanded neighborhood. Still, the public-private
partnership between San Francisco and Lennar was a far cry from the
nonprofit-sponsored affordable housing developments envisioned by the
Bayview-Hunters Point CDC and JHC in the 1960s. In fact, the toxic military base represented an emerging market for Lennar—a form of what, in
a different context, geographer Leigh Johnson has called "accumulation
by degradation."[6] During the 1990s and 2000s, the company also sought
development rights for four other contaminated, decommissioned military bases in the San Francisco Bay Area: Naval Station Treasure Island,
Mare Island Naval Shipyard, Alameda Air Naval Station, and the Concord

Naval Weapons Station.[7] Each military base required extensive environmental remediation. Yet in the context of a booming Bay Area real estate
market—and with the navy paying for the remediation work—Lennar was
poised to capitalize on a steep rent gap by converting these former bases
into upscale housing, along with offices and parks. The neoliberal form of
urbanization adopted by Brown and the SFRA—such as tethering social
goals, like urban redevelopment, to the profit motivations embedded in
public-private partnerships—arguably elevated the interests of corporate
shareholders over those of neighborhood residents.[8]

Longtime Black residents I spoke with during my fieldwork in the
2010s were unimpressed with Lennar's redevelopment plan for the shipyard, which by that time had expanded to include the adjacent Candlestick Stadium property, and promised over ten thousand housing units
and millions of square feet of office space. They would have preferred the
city focus on creating economic opportunities for current residents and
improving the neighborhood's existing physical infrastructure instead.[9] In
fact, the SFRA had initiated a neighborhood-scale redevelopment project in Bayview-Hunters Point in 2006, which was separate from Lennar's
project on the shipyard. The SFRA project (which, as noted in the book's
introduction, was opposed by some residents, who feared displacement
similar to what happened in the Fillmore) included the promise of local
jobs. Then, in January 2011, California Governor Jerry Brown announced
the elimination of all municipal redevelopment agencies in the state. At a
Redevelopment Project Area Committee (PAC) meeting at an elementary
school on Hunters Point hill I attended that same month, the executive
director of the SFRA called this a "dire scenario" for the agency's work
in the neighborhood, including the benefits promised to local residents.[10]
Although Brown's action would affect municipalities across the state, residents at the PAC meeting could not help but experience the dissolution of
the SFRA as part of a history of failed promises to their neighborhood specifically. During the discussion period, Patricia, who we met in the book's
introduction, said to the SFRA director: "You should be ashamed of your-
self. Think back to 1968 and redevelopment then. We're always last on the
totem pole. The only time we're first is on the picket line." Her comment
implied that state agencies could not be trusted, and that the only way
things got done in Bayview-Hunters Point was when residents organized
themselves.

Jackie had nuanced feelings about the SFRA neighborhood redevelopment project, although she shared a general lack of interest in Lennar's housing project at the shipyard.[11] I met Jackie while volunteering for Quesada Gardens Initiative, a grassroots gardening organization located in the Bayview. Quesada Gardens started with two elderly Black residents, Karl Paige and Annette Smith, who lived on a block of Quesada with a large median strip running down the middle of the street. In the 1990s and early 2000s, the median strip was used as an informal dumping ground. Residents of Quesada Street I spoke with during my time volunteering at the garden described how bags of trash, appliances, and car parts would appear on the median strip overnight, and how the San Francisco Department of Public Works (DPW) rarely cleaned up the site. In 2002 Karl and Annette began planting flowers on the median. Their neighbors filled buckets of water to help out with the growing garden and started planting flowers there too. They called DPW and other city agencies again and again, trying to get the trash picked up and to gain full access to the site, as a community garden project. On a Saturday morning in 2012 I sat with Annette on a porch step, watching volunteers prune trees and pull weeds on the median, which had become a well-tended garden of flowering trees, bushes, and other plants. Annette described herself as a farmer's daughter and told me she and Karl had eventually planted vegetables, such as collards and sweet potatoes. By the time I started volunteering, Quesada Gardens was a nonprofit organization and had converted three other vacant lots in the Bayview into gardens and other green spaces.

I had been dragging trash bins to the street after a day spent clearing out a weedy, trash-filled lot above a commuter rail tunnel—Quesada Gardens' most recent acquisition—when Jackie pulled over in her minivan and rolled down the window. "Are you going to plant food out here?" she asked me. "Isn't this soil toxic?" The organization didn't intend to seed edible plants there, but we spoke for a few minutes about the health concerns inherent in growing gardens in historically industrialized neighborhoods. I told Jackie about my research project on environmental politics in Bayview-Hunters Point, and we met up at the Southeast Community Center, just off Third Street, a few months later for an interview. Jackie told me that her parents had moved to San Francisco in the 1950s—she was proud that her dad had been a welder on the shipyard—and she grew up in Bayview-Hunters Point. As a young child she had lived on Navy Road, near the shipyard, before her family moved across Third Street to the Bayview. I asked Jackie for her thoughts on SFRA's redevelopment project in Bayview-Hunters Point. She had supported the SFRA at first—she thought the redevelopment plan looked good and liked that it would create jobs. After the financial and mortgage crises of 2008 and the dissolution of the SFRA, however, she was less hopeful. The SFRA's ambitious plans, and particularly the benefits for neighborhood residents, didn't seem feasible anymore. When I asked Jackie what she thought about Lennar's redevelopment project on the shipyard, she shook her head. "We don't need any more housing here, what do we need that for? Folks need jobs." Still, she added, emphatically, "they definitely need to clean it [the shipyard] up."

Like Jackie, the residents I interviewed for this project felt strongly that, whatever its reuse plan, the military base needed to be cleaned up, or remediated—a perspective reinforced by the archival record and from what I witnessed at meetings on environmental remediation I attended between 2011 and 2013. And yet while the navy and state regulatory agencies approached remediation as a technical project of reducing and managing toxic risk, and ultimately about meeting the conditions for the transfer of the property to the city (and from there, to the development company), most residents had a much different take. Remediation, many believed, ought to represent a form of reparative justice for the historical wastelanding of Bayview-Hunters Point and for the myriad effects of military abandonment. Reparative justice, in this case, involved the redistribution of economic benefits to the community (primarily in the form of jobs); investigating the potential health effects of the military base on residents and workers; and a formal, meaningful role for neighborhood residents in providing oversight of and influence on the remediation process.[12] Residents also insisted that aspects of the navy's remediation plan for the shipyard constituted an unacceptable risk to them and future generations.

Yet these expectations and desires conflicted with the science and policy of environmental remediation. For one thing, in contrast to the navy and state agencies, residents did not see the shipyard as an isolated property, separate from the broader neighborhood's history of extraction, abandonment, and toxic exposure. For another, although *remediation* is often used interchangeably with the word *cleanup*, remediation is not environmental restoration as it is typically understood. That is, remediation does not produce uncontaminated landscapes. Rather, Superfund sites are considered remediated once federal risk levels have been met.[13] In practice, this means that some amount of hazardous waste is often left on site and managed by remediation technologies such as landfill covers and land use deeds, along with promises for continuing monitoring and maintenance.[14] The navy's remediation project at the Hunters Point Shipyard involved a mixture of contaminant removal actions (removing contaminated soil, groundwater, and buildings from the site) along with other technologies to manage remaining toxic waste—in some cases, as was planned for a particularly controversial area of the shipyard called Parcel E-2, in perpetuity. Many residents saw the navy's plans to leave hazardous waste at the shipyard, and to call this "cleanup," as a continuation of the neighborhood's history of racialized state abandonment and toxic exposure. Indeed, the navy's remediation project eventually came to represent a form of environmental injustice.

Most broadly, this chapter looks at what it means to remediate, or try to remediate, a contaminated military base surrounded by a neighborhood that has lived with toxic exposure and state abandonment for over half a century. Bayview-Hunters Point residents suffer higher rates of environmentally related diseases than people living elsewhere in the city, and they have organized for decades to push state health and regulatory agencies to curb the excesses of industrial and military pollution in their neighborhood.[15] Because of these lived experiences, many residents see remediation at the shipyard within a historical, geographical, and ethical context that exceeds the scope of environmental policy and state regulatory agencies. Their critiques of navy-led remediation at Hunters Point offer one vision of what remediation centered on economic and environmental justice might look like. Yet their decades-long struggle on this issue also raises broader questions about what it means to repair a place when some harms, both social and ecological, are irreparable.

FROM PUBLIC HEALTH CRISIS TO ECONOMIC OPPORTUNITY

To understand the regulatory science underlying the navy's remediation plans for the Hunters Point Shipyard, we first need to understand historical shifts in U.S. policy on contaminated land. This regulatory history also deepens our understanding of why a corporate housing developer such as Lennar came to see a toxic military base as an economic opportunity and clarifies how and why *cleanup* does not mean *uncontaminated*.

In 1980, in response to a series of devastating toxic events—most notably, the upwelling of industrial chemicals into homes and a school in Love Canal, New York—U.S. Congress passed the Comprehensive Environmental Response, Compensation, and Liabilities Act (CERCLA), commonly known as the Superfund Act. CERCLA established a legal mandate and regulatory process for environmental cleanup of abandoned or uncontrolled hazardous waste sites. The law made polluting actors financially liable for cleanup and established a special trust fund (hence the name "Superfund") financed by crude petroleum and chemical feedstock taxes to pay for cleanup when no liable actors could be found.[16]

Passed by a lame duck Congress on the eve of Ronald Reagan's presidential inauguration, CERCLA was immediately inherited by an administration hostile to the very notion of federal environmental regulation. In his first term in office, Reagan oversaw sweeping budget and personnel cuts at the EPA, and federal enforcement of environmental regulations declined sharply.[17] His administration also appointed an industry lawyer, Rita Lavelle, to run the Superfund program. Within two years, Lavelle was indicted (and later imprisoned) for perjuring herself during a congressional hearing on corruption at the EPA.[18] The Superfund program had a halting start, and opposition to it expanded throughout the 1980s, with criticisms focusing on lax or uneven enforcement of the law and the perception that remediation projects took too long to complete.[19] By the 1990s, Superfund had also become unpopular with municipal governments, who blamed the law's environmental standards and strict liability clause for prolonging cleanups and disincentivizing land reuse (although lawsuits seeking to avoid corporate liability also prolonged cleanup).[20] Deindustrialization in the second half of the twentieth century had produced a growing landscape of contaminated, abandoned industrial sites across the country. Local officials worried about these sites depressing surrounding property values, eroding local tax bases, and creating public health risks. CERCLA deterred market actors away from Superfund sites; at the same time the decline of federal support for cities, beginning in the early 1970s, left municipal governments increasingly dependent on private capital for urban development projects. That is, the shift from what David Harvey calls "urban managerialism," supported by 1960s-era federal funding for cities, to "urban entrepreneurialism," or the reorientation of municipal governments around private sources of funding (as a response to reduced federal support and a decline in local tax revenues) meant that cities had fewer options in addressing abandoned, industrialized land.[21]

In response to these criticisms of Superfund, the EPA piloted its first brownfields program in 1993. The program introduced a new regulatory category, *brownfields*, later defined as a "property, the expansion, redevelopment, or reuse of which may be complicated by the presence or potential presence of a hazardous substance, pollutant, or contaminant."[22] In contrast to CERCLA, the EPA's brownfields program was, from the beginning, explicitly oriented toward private redevelopment as the solution to contaminated land. In this, the brownfields program reflected the agency's shift to a neoliberal approach to environmental policy—for example, by favoring market incentives such as relaxed cleanup standards, liability exemptions, tax abatements, and voluntary (rather than mandated) cleanup projects to attract real estate developers and investors to contaminated sites and to portray them as economic opportunities.[23] The brownfields program became widely popular, and it succeeded in stimulating market interest in contaminated land. In 1998, for example, the Urban Land Institute published the report *Turning Brownfields into Greenbacks: Developing and Financing Environmentally Contaminated Real Estate*, arguing that "it is now possible to reap substantial profits from . . . contaminated lands."[24] That same year, EPA administrator Carol Browner described brownfields as places of "opportunity," a term also used by the U.S. Congress in its 2005 hearings on brownfields, "Lands of Lost Opportunity."[25] Following the success of the brownfields program, in 1999, the EPA established the Superfund Redevelopment Initiative, to encourage and facilitate the redevelopment of Superfund sites as well. These policy shifts laid the basis for companies like Lennar to see economic potential in environmentally degraded sites.

In the context of the EPA's neoliberal reorientation toward the interests of market actors as a way of achieving environmental policy goals, the agency also revisited the practice of remediation. The implicit goal of CERCLA, when it was passed in 1980, had been "total environmental restoration."[26] Although cleanup standards are not established by law, remedial goals at Superfund sites were initially set to the highest environmental health standard, that of residential land use.[27] At the same time as the development of brownfields programs in the 1990s, however, the EPA adopted a risk-based approach to remediation that links baseline risk assessments with future land use or redevelopment plans.[28] Doing so, according to this logic, allows for quicker, less costly remediation projects, because—depending on future land use—much less remedial work is necessary.[29] For example, within a risk-based remediation framework, exposure risk levels for a future park or commercial district are lower than exposure risk levels for a future residential or agricultural site. The difference in risk is based on the estimated time an "average" person spends in a particular kind of place and the likelihood of human contact with different contaminants in different land use scenarios: people spend more time in their homes than in parks, for example, and engage in different kinds of behavior, while agricultural land uses (including gardening) can lead to bioaccumulation of contaminants through the food chain.[30] Ironically, as Shiloh Krupar has pointed out, remediation projects that result in parks or other apparently "natural" spaces are therefore likely to be the most contaminated underground.[31] To be clear, the practice of linking remediation goals with future land use was not simply about reducing cleanup costs. As a hydrologist at a state environmental agency, who had been working on toxic groundwater issues since the early 1980s, recollected to me in a white board-lined conference room at the agency's offices in 2012: "We [scientists] were scared in the eighties, about not being able to return contaminated land to its pristine condition. Over time I think we have realized it was not possible to clean up everything to "background," it takes so much time, it's just not very doable, and so over time we started to take this approach called risk-based cleanup."[32] A few moments later he clarified that risk-based cleanup "says, ok, there is some risk, but it is low enough." In one sense, linking risk assessments with future land use and aiming for residual risks that are "low enough" rather than nonexistent represented an acknowledgment of the impossibility of restoring landscapes and ecologies to preindustrial ecological conditions (although "low enough" must be understood as a sociopolitical and contingent rather than strictly scientific determination, and one in which "safe is synonymous with reasonable harm").[33] Still, these changes in the practice of remediation certainly aided the EPA's goal of attracting market interests to contaminated sites. As engineers Stephen Washburn and Kristen Edelmann explain in their history of risk-based remediation strategies: "Risk assessment has emerged as an important tool in determining the level of remediation necessary to return such [contaminated] sites to productive use, without requiring cleanups that make the property unattractive to developers."[34]

In line with this approach, aspects of the remediation of the Hunters Point Shipyard are structured to accommodate development interests— they are not, strictly speaking, about protecting human health. For example, instead of transferring the five-hundred-acre shipyard to the city as a single property, the military base has been divided into eleven distinct remedial parcels.[35] Each parcel moves through the Superfund process independently—that is, each parcel is governed by a distinct set of studies, remedial actions based on specific future land uses, and timelines to completion. When a parcel achieves its remedial goals, it can be transferred to the city, and then to Lennar, for development. This disaggregated, piecemeal remediation process allows the company to develop and sell properties on the military base and begin to see returns on its investment, even as the navy's remediation work continues on other areas of the base. In 2006, the shipyard's Parcel A was signed off by regulatory agencies as having achieved its remedial goals and was transferred to Lennar to begin its first phase of development, even though the Hunters Point Restoration Advisory Board (discussed later in this chapter) opposed the transfer, citing an insufficient cleanup effort.[36] By 2018, four hundred people were living in new townhomes at the shipyard, while remediation efforts continued around them.

SITE AND OVERSIGHT

During the course of my fieldwork, residents were especially concerned with a particular area of the shipyard, Parcel E-2. Parcel E-2 is a horseshoe-shaped, forty-seven-acre site on the southern portion of the military base and the location of four unlined landfills where the navy dumped hazardous waste during the 1950s, including radioactive materials used in shipbuilding and sandblast from attempts to decontaminate radioactive ships used in nuclear weapons tests in the Pacific Ocean.[37] Since the navy operated the shipyard prior to EPA regulations on toxic waste disposal, Parcel E-2's landfills were constructed and used without what are today considered basic public health measures, such as bottom liners, leachate systems, and landfill covers. I came close to Parcel E-2 several times on navy bus tours of the shipyard in 2011 and 2012. The site was always fenced off, with bright yellow signs marked with the triangular symbol of radio-activity and multiple warning signs in English, Spanish, and Cantonese, instructing people, in no uncertain terms, to stay away. As of the time of writing, the navy's remediation plans include leaving Parcel E-2's landfills on site and releasing the parcel for recreational use—as we will see, this has been a point of great contention for many Bayview-Hunters Point residents. This particular "remedy" for Parcel E-2 relies on two remediation technologies that have accompanied the institutionalization of risk-based remediation. *Engineering controls* are physical barriers such as concrete walkways, asphalt parking lots, or landfill liners, which aim to block specific exposure pathways between contaminants and humans. On Parcel E-2, engineering controls include industrial grade landfill liners covered by six vertical feet of imported soil. *Institutional controls* are administrative and legal mechanisms that manage human behavior, such as restrictive covenants that prohibit specific land uses.[38] On Parcel E-2, this includes prohibitions on building residences or schools on the site and on growing fruits or vegetables directly in the ground.[39]

Regulatory scientists I spoke with, in the 2010s, expressed great confidence that the remediation plans for Hunters Point were safe. Many pointed, in particular, to Superfund's multiple layers of regulatory oversight. To comply with CERCLA, the navy must submit to regulation by the EPA and two state-level environmental agencies, the Department of Toxic Substances Control (DTSC) and the Regional Water Quality Control Board (RWQCB).[40] CERCLA also outlines specific procedures, which include studying the nature and extent of contamination, evaluating different remedial actions, and justifying remedial decisions according to specific criteria—all elaborated within publicly available, if lengthy and difficult to read, documents. When I interviewed Andrew, a senior hazardous waste scientist from DTSC, in July 2012, he validated Bayview-Hunters Point residents' concerns about leaving the hazardous waste landfills on Parcel E-2, telling me: "I can understand. If I was in the community this would matter. I wouldn't want that stuff out there." "Still," he continued, "these are decisions the regulatory community must make." Andrew asserted himself as representing objective and universal interests, in contrast to a perceived particularism of Bayview-Hunters Point residents.[41] He felt that residents should trust the experts overseeing the Superfund process, assuring me, "There's a huge amount of eyes that see all the pieces, all the documents. If I was someone in the community, that would bring me a lot of comfort." A few months later I sat outside the RWQCB's offices in Oakland, drinking coffee with Jeffrey, a hydrologist with the agency.[42] I showed him a map of the shipyard, curious about whether groundwater contamination could travel from a parcel still undergoing remediation to a "clean" parcel. Jeffrey said that groundwater migration between parcels was possible, but unlikely; plus, he said, "the regulatory agencies, trust me, they're all over it." He continued, emphasizing this point: "I was about to make a joke the other day, about how many engineers does it take to figure out where to place a monitor. We had, like, fifteen people standing around these maps on a board, and I thought, this is just ridiculous." Jeffrey emphasized a surplus of oversight, similar to Andrew's description of a "huge amount of eyes." Both comments depicted regulatory oversight as all-seeing, implying that nothing would go wrong, and that Bayview-Hunters Point residents should trust the experts to do their jobs. As I discuss later, this confidence in regulatory oversight was not shared by the residents who showed up at remediation meetings. Yet even technical and legal documents included stipulations regarding the potential for toxic accidents, such as the pollution legal liability insurance taken out by the city of San Francisco to cover the costs of unexpected toxic events or discoveries at the shipyard.[43] In the world of Superfund, hazardous waste was simultaneously there and not there—in the case of Parcel E-2, the landfills would remain underground even as regulatory scientists told me not to worry about them. Or, hazardous waste would disappear in the development company's projections for the future of the shipyard only to pop back up in the fine print of insurance policies. Dr. Ahimsa Porter Sumchai, who grew up in southeast San Francisco and has been deeply involved in community oversight at the shipyard, characterized remediation as a "shell game" when I spoke with her in 2017. She explained, "If you don't keep your eyes on it, it disappears." I understood Dr. Sumchai's "eyes" as representing an objectivity tethered, or accountable, to the knowledge and experiences of Bayview-Hunters Point residents, and which refused to marginalize these forms of knowledge and experiences to the seemingly universalist perspective of regulatory agencies. Those eyes were attuned to the historical ways scientific reports and professional experts have failed neighborhood residents, and were much less convinced by the capacities of state regulation and technologies to manage and contain toxic waste on the shipyard.

DEBATING THE MEANING OF *CLEANUP*

On an afternoon in January 2016, Marie Harrison opened up the Bayview-Hunters Point Environmental Justice Task Force's monthly meeting and introduced Kay, a scientist from one of the environmental regulatory agencies that monitors the navy's project at the shipyard. The task force had been assembled by Greenaction for Health and Environmental Justice in 2015 and brought together state environmental agencies with local residents to discuss environmental problems in the neighborhood. The agenda for each meeting was informed by residents' submissions to an online, crowd-sourced map of environmental problems, which Greenaction maintained.[44] Agenda items during the years I volunteered with the mapping and task force project included idling diesel trucks, illegally dumped piles of trash in the neighborhood, raw sewage smells, and dust from the shipyard redevelopment work. Residents who showed up to task force meetings saw these problems as stemming from the state's failure to enforce environmental laws and regulations in Bayview-Hunters Point and interpreted these failures as systemic—part of the historical racialization of the neighborhood. The monthly task force meetings provided one way to try to hold state and municipal agencies accountable.

The January meeting focused on the potential impact of sea level rise on hazardous waste at the shipyard and included a presentation on the Superfund process by Kay.[45] The meeting took place at the Southeast Community Center in a spacious multipurpose room whose walls were lined with framed photographs of generations of Black community leaders. Kay's presentation walked the audience through each stage of the Superfund the process, emphasizing regulatory oversight at each step of the way. When Kay reached the final phase of remediation, "Closure and Base Transfer," she showed a slide depicting an aerial view of Lennar's future development, with colorful squares representing townhomes, office buildings, and parks. A central goal of remediation, this last slide indicated, was not simply cleanup but property transfer and redevelopment.

The slides communicated a seamless transition from contaminated military base to ordinary cityscape, but neighborhood residents in the room were not convinced. Gena, a Black woman who lived in public housing uphill from the shipyard, interrupted Kay, pointing to the "Closure and Base Transfer" slide.[46] "Is there still contamination, where all those houses are?" Gena asked. Kay hesitated, appearing to search for the right words. From the regulatory agency's perspective, the presence of contamination was not a public health problem, as long as engineering and institutional controls reduced the risk of toxic exposure to specific exposure levels. "No," Kay answered slowly, but then qualified her answer. "Not above action levels."[47] This sounded contradictory to Gena, who pressed the issue: "But is there still contamination? It's a yes or no question." Kay again hesitated, but then, redirecting her answer toward the question of safety, from a risk-based logic, assured Gena that the shipyard will "not be dangerous, according to our best science." Kay followed up this claim by detailing the engineering controls that would protect Lennar's development against sea level rise, including landfill caps and sea walls, along with the institutional controls: "no digging, no growing of vegetables, and long-term monitoring by the Navy, to make sure that cleanup is still in place." Gena responded to Kay, "But you said it was safe. Planting vegetables is something I should be able to do if the ground is safe. What if I want to plant tomatoes in the soil with my granddaughter?"

Kay's comment that the navy's post-cleanup monitoring of the shipyard would "make sure cleanup is *still in place*" indicates how remediation is a provisional accomplishment. Rather than a final, end state, remediation at the shipyard is a complex arrangement of things (landfill caps, sea walls) and the relations between people and things (no digging, no planting tomatoes), all of which must function appropriately to achieve compliance with specific exposure risk levels.[48] Remediation also relies on assumptions about the capacity of multiple remedial actions, such as landfill caps, to "hold" in the context of unpredictable climate changes, such as sea level rise; on the institutions responsible for monitoring the shipyard (in this case, the navy) following through with post-cleanup monitoring and other protocols; and on the compliance of future humans to abide by all land use restrictions. Achieving cleanup at the Hunters Point Shipyard—as geographer Shannon Cram has observed in the case of remediation at the Hanford Nuclear Reservation in Washington—depends on waste containment strategies and the production of reliable land users who will abide by particular behaviors.[49]

Gena's response to Kay's presentation proposed a different definition of cleanup and safety—stemming from her expectation that "cleanup" would allow for relationships, both among people and with the land, unmitigated by toxic risks. If the shipyard was truly cleaned up, Gena suggested, then she should be able to plant tomatoes in the soil with her granddaughter without having to worry about exposure to military waste. This image of gardening and intergenerational care and relationality functioned as a counterpoint to the state agency's definition of remediation, which depended on all the variables Kay had just listed.

From the perspective of state scientists, Gena's standards could be seen as uniformed about the logic of risk, requiring better science communication strategies from regulatory agencies and the navy. From another perspective, Gena could be seen as desiring the impossible—as M. Murphy writes, we live in a world so saturated by chemicals that industrial pollution has become part of "the molecular fabric of our bodies." As Murphy and others have pointed out, we need an environmental politics that moves beyond the notion that there is any "outside" or escape from industrial toxicity while also moving away from the techno-scientific habits, such as the state's risk-based logic, that have landed us here, "confronting the ubiquitous condition of chemically-altered human being," in the first place.[50] Still, Gena could also be interpreted as critiquing the state's constant use of the word *cleanup* to talk about remediation. As a man at the meeting later suggested, Kay was talking about "containment, not cleanup." I also think Gena's standards represented a kind of refusal. I first met her a few months before the task force meeting, when I presented Greenaction's online mapping project to a group of residents at a community center in the recently redeveloped Hunters View housing project. During the meeting Gena, who lives in one of the new Hunters View units, listed a number of concerns she had about toxicity in the neighborhood, including dust from Lennar's redevelopment work at the shipyard. We were sitting next to each other, and she swiveled around in her chair, pointing to a spot on the back of her neck. She could feel a lump growing there, and it worried her. Gena connected the lump with living in Bayview-Hunters Point and to the toxic environmental concerns she had just enumerated. Her expectations for cleanup at the shipyard, expressed a few weeks later at the task force meeting, can be understood as a refusal of the historical accumulation of toxicity in Bayview-Hunters Point and a broader disregard of the people who live there. It seemed to me that Gena knew that toxicity was as much about "wayward" chemicals as about systems of power, and she wanted something else from remediation than the reconsolidation of those systems.[51] Moreover, unlike state regulators and the navy, Gena did not see the shipyard as an isolated property or a set of distinct remedial parcels. Rather, the shipyard was part of a broader landscape of environmental harm in Bayview-Hunters Point. In this sense she was also introducing a different, historically and geographically situated context for evaluating remediation.

As I saw it, the difference between Kay's and Gena's standards for *cleanup* was not a relativistic point (merely that their perspectives were different). Nor am I trying to make comment about whether one definition is right or wrong. The point, I think, is about power and vulnerability—who has the power to decide what happens at the shipyard and who does not; who has had the power to determine land use and development in Bayview-Hunters Point more broadly, and who has not; and who has, historically and in the current moment, been subjected to these decisions and scales of governance. As residents struggled with these and other power imbalances and sought, over decades, to broaden the meaning of remediation at the shipyard, they also came up against the navy, which employed various strategies to maintain remediation as a strictly technical process.

REPARATIVE REMEDIATION

Since the early 1990s, when the navy announced it would clean up and transfer the Hunters Point Shipyard back to the city of San Francisco, Bayview-Hunters Point residents have sought to influence this process— to have a say in what remediation means and who the land is remediated for. Their expectations for community control over at least some aspects of the process took shape primarily through the Hunters Point RAB, which was established in 1993 at the request of neighborhood residents.

The stated goals of the RAB program—which is available to communities living near federal properties undergoing remediation, including military bases—include increasing public participation and serving as a "forum for the exchange of information and partnership" among local residents, the EPA, and the navy.[52] Organizationally, a RAB has an elected community board, including a community member who co-chairs the RAB with a (nonelected) federal representative. In the case of military bases, this person is the Base Realignment and Closure (BRAC) Act officer for that particular facility. Hunters Point RAB meeting attendees varied but often included staffers from state environmental agencies, local environmental and job placement organizations, private contractors working at the shipyard, and neighborhood residents.

In the 1990s and 2000s there was a strong showing of residents at RAB meetings and active participation on the RAB board. In general, residents who attended RAB meetings advanced an expansive vision of remediation that included acknowledgment of and accountability for past and present harms linked to the military base and its environmental and economic afterlives. Dr. Sumchai, who founded the Hunters Point RAB Radiological Subcommittee in 2001, remembered the RAB as a place for "fierce debates with representatives of the U.S. Navy, the Environmental Protection Agency and other state and federal regulatory agencies." The RAB had a "volatile history of advocating on behalf of the environmental health, public safety and economic development of the Bayview community."[53] Rather than acknowledge "fierce debates" as an aspect of public deliberation and dialogue—and perhaps par for the course in a neighborhood made and unmade through cycles of military mobilization and demobilization—the navy periodically threatened to dissolve the Hunters Point RAB altogether. In an interview in 2015, a former RAB member also recollected how meetings in the 2000s were held "in the police station, not the community." In a neighborhood with a history of anti-Black police violence, holding meetings in the police station would have functioned as an intimidating, exclusionary tactic.

Navy BRAC officers cochairing the RAB regularly reminded Bayview-Hunters Point residents that the military was only responsible for the "environmental" side of the shipyard redevelopment project, which they defined as the discrete, technical tasks required to achieve compliance with risk-based remediation goals and transfer the property to the city. For residents, however, the question of how many truckloads of toxic soil had been removed from the shipyard each month was inseparable from the question of who was getting paid to drive those trucks. Were Southeast residents being put to work? The lack of jobs and the demand that the navy take up the responsibility for providing them was a consistent topic at RAB meetings. As part of the RAB agenda, the navy would present updates (*this* number of surveys conducted, *this* amount of soil removed), whereas residents wanted to know who, exactly, benefited from all that progress. Consider the following exchange from a meeting on July 27, 2000. The navy's lead remedial project manager had just explained recent soil sampling efforts on Parcel B and then asked if there were any questions.

MS. DOROTHY PETERSON [Hunters Point resident and cochair of the RAB]: How many Bayview-Hunters Point residents are you using now?

MR. DEMARS [lead project manager for the navy]: I don't know. I could defer that question to Jim.

MR. JIM ROBBINS [IT Corporation]: Right now we are in the process of interviewing and hiring local residents. We have issued subcontracts to a couple of firms already, and we are still early in our process.

ATTENDEE: So all this work that's getting done, you have nothing to show to help this community?

MR. ROBBINS: I don't think that's quite true. We have got A. M. Kayans, a local firm we have been using to haul in some of the material we need to build decontamination pads.

ATTENDEE: We need figures.

MR. ROBBINS: You want a dollar figure?

ATTENDEE: No. Number of people.

MR. ROBBINS: We have not hired anyone to date and put them on the work on the site yet.

The conversation continued. A certain percentage of remediation jobs (mostly temporary, manual labor) had been promised to Bayview-Hunters Point residents. The fact that they had not yet appeared, said Dorothy Peterson, meant that the navy and its subcontracting company were not fulfilling their responsibilities. She said to Mr. Robbins: "So I think you really need to correct that and correct it immediately." She then offered the phone number for Jesse Mason, another longtime resident and former shipyard worker, who sat on a different advisory committee having to do with redevelopment on the shipyard, telling Mr. Robbins that he could call Mason to make those hires.

Ms. Peterson's demands could be understood as a narrow form of distributional politics, which is how naval officers often seemed to treat local residents' demands. For example, at a remediation meeting I attended at the Bayview-Hunters Point YMCA in December 2011, a BRAC officer assured the audience that "this area will get a piece of the piece." However, this interpretive framework misses the historically extractive relationship between the military and Black residents in Bayview-Hunters Point, as well as the ways the neighborhood was left out of San Francisco's postwar economic boom. Ms. Peterson's demands are better understood within the neighborhood's long history of advocating for economic sufficiency. At the same RAB meeting, for example, she also expressed concerns about whether Bayview-Hunters Point remediation workers could transfer to other base cleanup projects once their work in Hunters Point was complete. Ms. Peterson was advocating for stable and permanent, rather than temporary, jobs for local residents.

RAB transcripts from the 2000s also reveal a sustained inquiry into the shipyard's toxic history and its potential health effects on former workers and local residents, with a particular focus on the secretive NRDL. Residents worried about radiation from the military base and nuclear laboratory and wanted the navy to take responsibility for investigating these potential harms. For example, in the summer of 2001, two experts from the navy's Radiological Affairs Support Office (RASO) in Virginia flew to San Francisco to answer a series of detailed questions about the NRDL. In an exchange at an RAB meeting on July 26, 2001, one of the RASO experts addressed a question posed by the RAB board: "Did the NRDL conduct human radiological experiments?" In 1944, in the context of developing the atomic bomb, the U.S. government initiated human radiation experiments on unwitting patients across the country. The Department of Energy declassified state documents on these experiments in 1994, a year after the Hunters Point RAB had formed (thus, it made sense that the RAB board would pose this question).[54] The RASO expert answered that yes, the NRDL did conduct radiation experiments on people, but she qualified her answer—that it was done only "under controlled conditions as authorized by an Atomic Energy Commission license." Everything was legitimate, or at least authorized by the state. The RASO expert tried to reassure the RAB that "Bayview residents were never involved in the human experimentation. All the human experimentation was done within the lab." Dr. Sumchai pushed back on this remark, challenging the RASO expert's time- and place-specific conception of "the lab" and expanding military's responsibilities to longtime residents in the neighborhood. Dr. Sumchai pointed out that the NRDL released radiation into the local environment. "If you released hazardous radioactive material into the air, water or soil and you admit it, then the people of this community were affected," she said. In this exchange are two different conceptions of the sociospatial boundaries of the NRDL, and by extension, two different geographies of responsibility.[55] For the miliary, the NRDL's experimentation on people involved discrete, temporally and spatially bounded studies. Experimentation took place within the four walls of a designated building and the temporal boundaries of an "experiment." Forms were signed, data collected, conclusions reached, and papers filed. Dr. Sumchai's point, in contrast, was that anyone living near the NRDL could be considered unwitting human subjects of radiation experimentation. She extended the sociospatial boundaries of the lab and therefore the boundaries of the navy's responsibility to longtime residents.

Percy, an active RAB member during the 2000s, was also worried about the spread of military radiation throughout the neighborhood.[56] His parents had moved to San Francisco during World War II—his mother came from Arkansas, his father from Louisiana—and they worked at two different Bay Area shipyards. Percy grew up in the Fillmore district before moving to Hunters Point along with other Black San Franciscans pushed out by urban renewal in the 1960s. Percy served on the board of his tenants' union in Hunters Point and had previously worked with environmental justice organizations. I spoke with him over the phone three times between 2010 and 2011. Some of his kids had asthma, he told me, and he connected their illnesses to vehicle emissions from the two freeways that run along the length of Bayview-Hunters Point and from "the dirt from the shipyard, from the trucks hauling out the stuff. It gets to them, like the fire from the landfill." In August 2000 a fire had smoldered underground on Parcel E-2 for nearly a month. Residents living uphill from the shipyard, including Percy, described seeing or hearing about puffs of yellowy-green smoke emanating from the ground. Yet it was not until September 1, a few weeks later, that the EPA took air samples at the site.[57] The landfill fire was a reminder, to many residents, of the "undead" nature of militarization and the failure of state health agencies to protect them.[58] The fire and its mysterious-colored smoke was often invoked at public meetings and events I attended in the 2010s as evidence of the underground dangers at the shipyard and the inability of the navy to completely understand, let alone control, its toxic legacy. Over the phone, Percy told me he had started organizing for environmental justice because "I didn't want my kids to die, so I started fighting."

Percy had a lot to say about contamination at the shipyard and was particularly concerned about radiation. During one phone call he told me, "The people who worked there [at the shipyard], they didn't know about the nuclear testing. They carried it home, on their clothes, shoes." By "it," Percy referred to radiation. Percy's concerns that shipyard workers brought radioactive materials from nuclear testing home with them were not unfounded. In January 1950 a glass capsule of radium sulfate broke open on the floor of a laboratory on Naval Station Treasure Island, in the San Francisco Bay. At the time, Treasure Island ran a training program in atomic, biological, and chemical defense. The spill exposed thirty-five crew members to dangerous levels of radiation and forced an emergency closure of the site. Radiation scientists from the NRDL in Hunters Point were called to Treasure Island to clean up the contamination, which was so extensive that—according to a study later published by NRDL researchers in the journal *Nucleonics*—their Geiger counter readings showed radiation levels off the scale.[59] In the words of one navy officer later interviewed for an article in the *Los Angeles Times*, some of the crew members carried so much radium sulfate home, "we could trace their footsteps with Geiger counters. We could tell whether they went into the living room and kissed their wives, or walked into the kitchen for a glass of water."[60]

Percy imagined a similar scene in Hunters Point. On the phone, he continued expressing his concerns about radiation, but switched to the first person and the present tense—a more urgent register. "I go home and my children are on the floor. I cross the rug and they roll in it. The NRDL, they left after thirty years, before the cancers." What Percy brings home to his children is qualified by his reference to the radiation laboratory— he brings home invisible radiation. Percy depicts an intimate relationship between shipyard and home, in the ways war's residues pass through the home's threshold. His anxieties about guarding his home, as a form of care for his children, represents a form of what geographers Pavithra Vasudevan and Sara Smith call *domestic geopolitics*, or the everyday strategies "engaged by domestic populations to defend their bodies, homes, and communities from the unspectacular, yet insistent, everyday onslaught from the imperial state"—in the case of Hunters Point, through the material and affective afterlives of preparing for nuclear war. For Vasudevan and Smith, domestic geopolitics names the "daily negotiations with death-dealing structures that complicate, and exceed, the already challenging demands of obtaining food [and] maintaining a household" for "communities of color living in capitalism's toxic 'sacrifice zones.'"[61] For Percy, part of this daily negotiation involved the desire to protect his children from the specter of invisible harms outside his apartment door.

Bernice was also concerned with accounting and accountability for toxic injury when I formally interviewed her at the offices of the nonprofit where she worked in October 2012. Bernice sat behind a stately wooden desk and recounted to me some of her experiences on the Hunters Point RAB. Bernice, who had moved to San Francisco from Missouri, began working at the Hunters Point Shipyard as a teenager, just shy of her eighteenth birthday. Similar to Percy, she first lived in the Fillmore district and eventually moved to Bayview-Hunters Point. Bernice was proud of her national service, while also deeply concerned with the navy's environmental legacy in Hunters Point. She told me:

> My first encounter with the Navy—over the contamination that was left, as a civilian at the shipyard—I asked the commanding officer, how many military families that lived there contracted some type of cancer, like mesothelioma? And he said to me, well I can't give you that, no one knows that kind of information.
>
> And I said, oh I beg to differ. I used to work for the Navy, did I tell you that? Did I also tell you I was a file clerk specialist? He looked at me and he said, what does that mean? And I said, it means that I couldn't order a box of pencils without it being signed off, like, three times. In my office, my commanding officer had to sign off for me to get an aspirin, because I had a headache. But in order to get that aspirin, I had to have the commanding officer sign off, then the doctor sign off, and then the pharmacist sign off. So that was three different signatures, on a form that was tripled, and they kept a copy, and the pharmacists kept a copy, and I brought a copy back to my commanding officer, that I had to file!

As a file clerk specialist, Bernice knew the navy kept meticulous records. She knew the navy monitored and counted what mattered to the navy. Her story satirized the military's near-clinical attention to detail, or at least to some details, while making clear that the navy was quite capable of accounting for workplace cancers. From the BRAC officer's perspective, however, it was obvious, and did not need to be explained, that the military would not have "that kind of information" and therefore could not be held accountable for lung cancers navy workers or family members developed later in life. In contrast, Bernice posited the military's ignorance as socially produced, a form of "undone science."[62]

Bernice also pushed back against the logic of environmental remediation, which she felt similarly functioned to circumscribe the military's responsibility to the longtime Bayview-Hunters Point community. She told me about a different remediation meeting, at which she challenged the BRAC officer on the topic of groundwater contamination. Bernice was concerned that groundwater from the shipyard was leaching into the bay. Hunters Point was a popular fishing spot, and anglers might be reeling in toxic catch. As she recounted to me, the BRAC officer told her the navy owns one hundred yards of water from the shipyard waterfront line, and that any contaminated groundwater leaching into the ocean would most likely fall within its ocean "property." Bernice responded to him: "I need to say something. I'm not a scientist, I'm not an oceanologist, what I am is a mother and a resident of this community. So that's what I am, and now let me tell you what I know." Setting up her own position as a mother in contrast to the role of professional engineer, she was elevating care and relationality as forms of relevant expertise. She went on:

> I know when the tide rolls out, and it comes back in, it doesn't just pass over everything that just leached off the land, into the water. I know that when the fish swim into your, quote, hundred yards of water that you own, they don't get up to this invisible line and say, that's the Navy's land! That water belongs to the Navy, and you know, I can't swim there because it's contaminated. A lot of people are catching fish and crabs and shrimp, which are bottom feeders. Now, are the crabs and shrimp as smart as the fish who get up to that invisible line and go, oops, can't go there!

Bernice's critique, threaded with her dry humor, seemed lost on the BRAC officer. He reminded her that she wasn't an engineer, and that she couldn't possibly understand the technical details of groundwater flow. This irritated Bernice. She took her cup of coffee and poured it into the BRAC officer's half full glass of water. "Oh damn," she said to him, "would you mind telling my coffee to get back in my cup, because I own it and you own the water, and by the way I want my coffee warm as it was when I poured it in there?" In this act of political theater, Bernice used her coffee to symbolize toxic pollutants, while her impossible demand to separate out the two liquids parodied the BRAC officer's logic—his insistence that contaminated groundwater from the shipyard could be functionally separated from the rest of the ocean. Her demonstrative unreasonable-ness highlighted the officer's logic as similarly unreasonable.[63]

Bayview-Hunters Point residents advocated for remediation as an expansive socioenvironmental project by advocating for local hiring policies, health studies, historical accountability, and by insisting on what, and whom, the RAB was "for." Additionally, Percy and Bernice showed how the shipyard and its toxic legacies are embodied political ecologies.[64] They expressed intimate experiences and anxieties, as caregivers, about environmental toxicity from the shipyard. For Percy, this was the image of his children playing on the floor; for Bernice it came from her role as a mother and concerns about the waterfront near her home, and for her neighbors who fished there. Bernice and Percy, however, had been able to express their frustrations and disagreements in RAB meetings. This formal channel was soon shut down.

THE NAVY DISBANDS THE RAB

From the navy's perspective, the Hunters Point RAB was unwieldy and inefficient. It consistently deviated from its narrowly defined purpose of "advising" the navy on remediation. Questions about radiological history or jobs were supposed to be separate from the specific, technical task of remediation and base transfer. The navy BRAC officer from the early 2000s, Richard Mach, often displayed his exasperation and condescension toward local RAB members when they deviated from this task. For example, at a meeting on September 29, 2001, Mach announced—to the RAB board's surprise—that the navy had hired an outside facilitator. Mach said, "There had been many, many complaints about the running of this RAB and the [...] degradation of it being run in a smooth manner, and we have gone and solicited for trained facilitators to run the RAB, and we have chosen one, and we are hiring that person."[65] Mach expressed frustration with what he perceived as the RAB's inefficiencies: "We have spent the last five months bickering about how to set up subcommittees [and] bickering about how they are going to be run." His choice of words is patronizing—children "bicker," and "bickering" is petty, insignificant speech. Moreover, he asserted control over the RAB unilaterally, without consulting the board's elected community members.

Bayview-Hunters Point residents and advocates at the meeting rejected Mach's actions as offensive and illegitimate. Dr. Raymond Tompkins, a Black scientist and college professor, who had grown up in military housing in Hunters Point, told Mach that his actions said "our opinion means less." Community organizer and RAB member Marie Harrison also admonished Mach: "I view what you did as . . . one more slap in the face to this community. No subcommittee has the authority without bringing it before the full board to make any such decisions." She continued, framing Mach's actions as a form of job theft: "Is it not part of the RAB agenda to keep contracts within the community, that we should hire from within the community when possible? And I believe that was possible, if I'm not mistaken.... [T]hat takes one more contract out of this community [and] that violates what you're here for."[66] Dorothy Peterson had also explained to Mach what the RAB was "here for" at a meeting in August 2000. As community co-chair, she requested the RAB put Lennar on the meeting agenda. The city had recently awarded Lennar development rights to the shipyard, yet the details of Lennar's redevelopment plans were unclear. Mach rejected this idea, telling Ms. Peterson "the RAB is for the environmental side," not for discussions about redevelopment. Dorothy Peterson immediately corrected him, "No, the RAB is also for the community. And the shipyard involves the community, and we want to know what's going on with this transfer."

Debates over the definition of the RAB, the meaning of remediation, and who the RAB was "for" continued throughout the 2000s. For the navy, the RAB was, at the end of the day, a box to check as part of its goal of transferring the shipyard property to the city, and ultimately to the development company, in a timely and efficient manner. In contrast, residents saw the RAB as a space for negotiation and accountability. Many residents came to RAB meetings with family histories of living next to and working on the military base. Through the RAB, they demanded recognition of and compensation for the toxic economic and environmental legacies of the shipyard. Still, the federal RAB program was not designed to meet these expectations and demands. In 2009, when the navy saw that the RAB had become ungovernable, it simply terminated it.

Its justification involved residents' mounting frustrations over Lennar's redevelopment project at the shipyard—which, unsurprisingly, became a topic at RAB meetings—particularly after Lennar started leveling the hillsides on the shipyard's Parcel A in 2006. I discuss residents' concerns over redevelopment dust, which included asbestos, generated by these grading activities in chapter 4. Here it is enough to say that anti-Lennar organizing brought new social actors into the center of shipyard redevelopment politics. Leaders of the Nation of Islam church in Hunters Point—which ran a school adjacent to Parcel A, separated by a chain-link fence— became outspoken critics of Lennar and the San Francisco Department of Public Health (SFDPH), which asserted that Lennar's redevelopment dust was safe.[67] The church began hosting Thursday "town hall meetings," which drew large crowds, and allied with other Bay Area organizations active in Bayview-Hunters Point, including People Organizing to Win Employment Rights (POWER), to form the Stop Lennar Action Movement (SLAM). One of POWER's organizers at this time was Alicia Schwartz, who later married and changed her last name to Garcia, and, in 2014, co-founded Black Lives Matter.[68]

During 2008, organized opposition to Lennar in Bayview-Hunters Point pursued several courses. First, POWER and its allies sought to make housing at the shipyard redevelopment project more affordable to people living in the neighborhood. POWER gathered enough signatures to place Proposition F on the June 2008 city ballot, requiring 50 percent of Lennar's new housing stock be affordable to people earning between 30 and 80 percent of the city's average median income (AMI). Lennar, in contrast, had promised 32 percent of its housing stock would be affordable to people earning 60 to 100 percent of the AMI—a much higher income bracket, and less accessible to many Bayview-Hunters Point residents. Moreover, Lennar was not legally bound to this number, whereas Proposition F made affordable housing provisions legally binding. As Alicia Garcia née Schwartz wrote in an op/ed in the *San Francisco Bay View*, a local paper, supporters of Proposition F believed that "another Bayview is not only possible, but NECESSARY to ensure the survival of African Americans and working-class families in our city."[69] Lennar spent $4 million on an anti-Proposition F campaign, and the ballot measure was defeated at the polls.[70]

Second, the SLAM coalition turned toward the RAB. Leon Mohammed, a leader of the Nation of Islam church in Hunters Point, was elected a community cochair of the RAB in January 2009, along with six other new RAB members. The new voting bloc proceeded to pass a resolution requesting a different representative from the SFDPH to the RAB. Many Bayview-Hunters Point residents, and particularly those organizing with SLAM, had lost trust in the current SFDPH representative, Amy Brownell, over her handling of the asbestos and dust issues during 2006–2007. A few years later, in 2011, documents received through a Freedom of Information Act (FOIA) request by a Bay Area paper, the *Bay Citizen*, revealed a close relationship between Brownell and Lennar. In one email to a Lennar employee on October 13, 2006, for example, Brownell wrote,

"I'm sure you will also want to change my wording on how I portray the problems, lack of monitors, etc. Go ahead and change any way you want. I may change some of it back but I'm willing to read your versions."[71] The emails suggested that the company had undue influence on SFDPH's policies. Although the RAB board did not have access to these documents in 2009, many Bayview-Hunters Point residents already felt that the SFDPH ignored their health concerns in favor of allowing Lennar's construction to continue; the FOIA-requested emails later merely reinforced this suspicion. Then, in February 2009 the newly empowered RAB board voted to replace the navy BRAC officer assigned to the RAB as well. In response, the navy dissolved the entire Hunters Point RAB program. In the letter of dissolution, the navy cited a "hostile tone" and the fact that "RAB meetings are used to discuss non-Navy issues" such as redevelopment and "contracting issues rather than a cleanup program."[72] These had been recurrent complaints by the navy about the Hunters Point RAB, but the new, oppositional voting bloc represented the last straw.

The navy replaced the Hunters Point RAB with a PowerPoint-driven presentation meeting format it called "community informational meetings." The format was not dialogic or deliberative, but a one-directional transmission of information. There were no decisions to be made or debates to be had at community informational meetings, only information to receive.[73] Between 2011 and 2013 I attended most of these monthly meetings, with an occasional return in subsequent years. Having read RAB transcripts from the previous decade, I could tell that local attendance had dwindled, most likely a result of the harsh dissolution of the RAB and the year and a half it took the navy to start up public remediation meetings again, in addition to the presentation-based, rather than discussion-based, format. As a former RAB member told me, "The important thing about the RAB being disbanded—there's no public input." Still, a few longtime RAB members continued to show up and demand a level of accountability the navy was unwilling to provide.

Longtime RAB members were critical of the new, unidirectional meeting format. At a community informational meeting in March 2011, for example, the navy BRAC coordinator delivered a presentation on recent progress at the shipyard and then separated the audience into small groups for a structured forty-five-minute breakout session. The BRAC officer and state environmental regulators stood in different corners of the multipurpose room in the Bayview Opera House while audience members were instructed to walk around in small groups and ask questions. This structured activity left very little time for open debate. Neither were conversations recorded and archived, as the RAB meetings had been. During the short public comment period, an organizer with POWER pointed this out, saying that the new meeting format was

> geared to one-way communication. There should be a formal democratically elected community body, even if they don't have decision making power, to record people's experiences, whether or not they agree with the Navy. As of now, there is no record, no formal place where people can record their concerns. What the RAB *did* do, it formally recorded if people disagreed with the Navy.

RAB meetings were transcribed, word for word, and most (though not all) archived in the San Francisco Public Library.[74] These transcripts provide a public record of the ways Bayview-Hunters Point residents disagreed with the navy and tried to redefine remediation through notions of environmental and economic justice, over many years.

Some of the exchanges I witnessed at community informational meetings felt devastating. For example, at a meeting in January 2011 several residents raised health concerns about construction dust continuing to come from the shipyard. Amara, a Black resident in the audience, asked, "How does an average citizen find out the daily exceedances [a term often used in remediation meetings to refer to air pollution generated by the navy or Lennar that exceeded state health air quality standards] from the shipyard construction? It's still killing people."[75] Amara expressed a feeling of embodied vulnerability, and—whether or not the navy agreed with her interpretation of events—her words ought to have become part of the public record. Yet as far as I know, the only record of her concerns is my handwritten notes from that evening. In contrast to the RAB's word-for-word transcripts, the only official records of community informational meetings are the navy's PowerPoint slides and glossy handouts, posted as PDFs to the navy's BRAC website. This one-sided textual record represents the shipyard remediation project as a story of slow and steady progress, rather than a deeply contentious process.

Other residents who showed up at community informational meetings continued to demand jobs and other economic benefits. In August 2011 I sat in a multipurpose room at the Asian Pacific American Community Center in Visitacion Valley, another neighborhood along Third Street in southeast San Francisco. The BRAC coordinator, Keith, had just finished his PowerPoint presentation on Parcel E-2. Evelyn, a longtime Filipinx organizer from Visitacion Valley, spoke up from her seat in the back of the room.[76] "What's the cost of E-2 to taxpayers?" Keith replied that remediation of Parcel E-2 would cost $70–$75 million, and that the total cost for cleaning up the shipyard was in the realm of $1.1 or $1.2 billion. An Asian American man spoke after Evelyn, asking Keith about the distribution of that money. "Does all this go to big corporations? What about neighborhood businesses?" Keith hedged his response, saying that the navy had twelve primary contractors, and each contractor had its own subcontractors. The navy had tried to hire graduates from City Build, a Bayview-Hunters Point–based job training organization, but "it's unfair to hold job fairs when there are no jobs." He further distanced the navy from any economic accountability to local residents, adding that it was not bound to San Francisco's local hiring policy, which did not apply to federal agencies. The man in the audience was unconvinced by Keith's bureaucratic maneuvering, and quipped, "We got the bad air, but we don't have the jobs." His terse comment underscored the uneven distribution of benefits and harms through the business of remediation and points to the ways remediation can be an extractive rather than reparative process. Amara made a similar point at the community informational meeting in January 2011. After raising her concerns about air quality exceedances, she told the BRAC officer: "The Navy has the capacity to hire more locals. Since the Navy is putting the community in danger, they should take some danger out of the community." The first danger she spoke of referred to air quality exceedances and respiratory health problems, while the second danger was the historical lack of jobs and economic hardship.

During the meeting at the Asian Pacific American Community Center, a white scientist with an environmental organization (which often took a critical stance on the navy's remediation work) spoke up. "Everyone is talking about jobs, jobs, jobs," the scientist said. "This is not a jobs meeting, this is an environmental meeting. People are frustrated because they come to environmental meetings to talk about jobs. People need to be aware that there are other ways, other people to talk about jobs with, but not at a cleanup meeting." The scientist sided with the navy's interest in maintaining remediation as a strictly technical process—in his view, the question of jobs was irrelevant to the point of that evening's meeting. After a moment of silence, Evelyn shot back, "With the release of $1.2 billion [the projected cost for remediation at Hunters Point], people have every right to ask about jobs." Evelyn rejected the scientist's discursive separation between jobs and the environment. Indeed, southeast San Francisco residents were quite clear that environmental remediation was a big business and these profits were reaped by people who lived outside the neighborhood. At a different remediation meeting in 2012, I heard a man say loudly, to a packed audience at the Bayview Opera House: "When I hear 'Superfund,' I know that someone's making a whoooole lot of money."

Both the Hunters Point RAB program and the navy's subsequent community informational meetings functioned as disciplinary spaces, producing legitimate and illegitimate topics of discussion. While it should be no surprise that the delimited space of the RAB would seek to capture or dilute oppositional politics, it is notable that residents persisted in trying to make the RAB work "for the community," as Dorothy Peterson put it. Their critiques and demands over the years focused on widening the state's technical work of remediation into a socially as well as environmentally reparative project, one that included acknowledgment for past harms, environmental health studies on the potential health impacts of the shipyard on local residents, economic redistribution, and a role in defining acceptable and unacceptable forms of toxic risk. The question of acceptable risk was most apparent in a widespread rejection of the navy's remediation plans for Parcel E-2.

"WE'VE LIVED WITH THIS STUFF FOR TOO LONG": PROTESTING THE LANDFILL CAP ON PARCEL E-2

I maneuvered my car into a parking spot a few blocks uphill from the Ruth Williams Memorial Theater, at the Bayview Opera House, and walked quickly down the sidewalk against a brisk wind, late to the meeting. It was a characteristically cold and gray San Francisco evening in July 2012. As I approached the theater, I heard the rise and fall of voices—the buzz of a large audience. The auditorium was packed with people mingling and slowly making their way to seats in folding metal chairs. I chatted with a few people I recognized and then found a seat on the left side of the room.

Three white, male scientists sat behind a long table at the front of the room. Each scientist represented a different professional discipline: toxicology, landfill engineering, and radiation science. They had been contracted by Arc Ecology, a military-environmental watchdog group that had its offices in Bayview-Hunters Point, to provide an independent evaluation of the navy's remedial plan for Parcel E-2. Their time and expertise was paid for by a Technical Assistance Grant (TAG) from the EPA, which provides funding for communities living near Superfund sites to understand the complexities of the science of remediation. At the start of the meeting, Arc Ecology's director, Saul Bloom, explained that this event was a "meet and greet," intended for the scientists to introduce themselves to Bayview-Hunters Point residents and hear about neighborhood concerns, before they undertook their evaluations of the landfill remedy. The scientists would report back their conclusions at another meeting at the Opera House in the following month. Bloom passed the microphone to the table of scientists, and each introduced himself, providing a verbal résumé of his professional expertise.

As the scientists spoke, Black Bayview-Hunters Point residents assembled behind a standing microphone, forming a long line that stretched down the aisle in the middle of auditorium. Donna, a lawyer, longtime resident of Bayview-Hunters Point, and former RAB member, was one of the first to speak.[77] "When I was born, my parents lived on the shipyard," she began. Donna raised her children in the neighborhood. She said, "This is an environment justice community, and the EPA is required to take this fact into consideration—to apply higher standards in their risk assessments. Even though we are no longer the majority, we are still *the* African American neighborhood in San Francisco. If you look at the public health research for the past thirty years, it shows that disparities in health for African Americans has gotten worse, and not better." Like others who spoke that evening, Donna instructed the scientists to include the history of racial segregation and related, place-based health disparities in their evaluation of Parcel E-2's landfill cap.[78] Moreover, for Donna and others at the meeting, the professional independence of the scientists—they were not on the navy's payroll or connected to any of the companies paid to do remediation work at the shipyard—did not necessarily translate into their having the authority to evaluate the best remedial action for Parcel E-2. Marie Harrison spoke after Donna, echoing this critique. "Instead of re-lying on the navy's science," Marie proposed, "come talk to someone out here, who has watched people suffer and die." Marie provincialized the state's science as one among many possible interpretive frameworks and broadened the kinds of evidence she thought the three scientists ought to rely on in their analyses.

Donna's and Marie's comments received applause, while the line in front of the microphone continued to grow. The size of the crowd and length of the public comment queue spoke to the significance of leaving Parcel E-2 landfills on the military base to Black residents in Bayview-Hunters Point. One middle-aged woman framed the landfill cap as part of a longer history of abandonment and the social production of embod-ied vulnerability in the neighborhood. She said, "I've lived here, across from this Superfund site, for forty years." The woman spoke about her father having served in the military, a point that emphasized her fami-ly's sacrifice to the nation. "We don't care what you want to build over there, people's health should come first. If your pocketbook comes up short, well, our pocketbook has been coming up short for years. I'm tired of people getting money while we suffer." Like many other residents I in-terviewed over the years, the woman was uninterested in redevelopment at the shipyard, but held clear opinions about what constituted a proper cleanup. She also characterized remediation as a racial capitalist proj-ect—distributing both wealth and environmental harm in ways that re-inforced existing racial-spatial formations. The woman was worried that the proposed landfill cap for Parcel E-2 would break, and like many in the audience that evening, she referred to the underground fire at Parcel E-2 in August 2000 to reinforce her point that toxic accidents at the ship-yard were possible. "You need to dig everything out," she insisted, adding that "concrete cracks—you walk outside, you can see concrete cracking all over the place." Although the landfill cap would be plastic, not concrete, her comment ought to be understood in relation to the long-standing forms of state infrastructural neglect in Bayview-Hunters Point and as indicative of a profound mistrust of state agencies to protect residents from environmental harm. The woman concluded, "We've been fighting this too long." Her use of the word *this* indicated how the toxic landfill represented a broader, systemic issue. Her dismissal of a cost-benefit analysis—which was part of the navy's justification for the landfill cap on E-2—was echoed by many others who spoke that night. "You can't put a price on our lives," said an elderly man. Willie Ratcliff, editor of the *San Francisco Bay View*, continued this line of thought: "How can you balance $300 million dollars [the cost difference between a landfill cap and full excavation of the landfills on Parcel E-2, according to technical remediation documents] against future generations?" He continued, "No way in hell we should let them keep E-2, I don't care what it costs." Another older man reiterated that the landfill in question was not an engineering problem but a part of lived, multigenerational experience of toxicity, which required historical and ethical rather than strictly technical considerations. "I have nine grandkids. We've lived with this stuff for too long. I've lived in Hunters Point, near the shipyard, my whole life. Whatever has to be done, whatever money needs to be spent, it's worth it."

Toward the end of the evening, Patricia stood up at the microphone. Born in Texas, Patricia had spent part of her childhood in public housing near the shipyard. I had interviewed Patricia a few months before the Parcel E-2 meeting. She told me her husband had been part of the "navy brass" and spoke of his national service with a sense of pride that communicated an older, more positive association with the military. That evening she said, "I don't have confidence in anyone anymore. In 2001, the government started sending ten billion a year to fight wars overseas. I asked our representatives, could we have a one-time ten billion to clean up the shipyard?" Patricia wanted the entire shipyard cleaned up to residential standards, which included "no capping." For Patricia and other residents, the fact that Parcel E-2 would only be safe for recreational purposes meant that the shipyard would not be remediated to the highest standards possible, which is what they felt the community deserved. Patricia also called attention to the huge sums of money allocated to the U.S. military's ongoing wars, in contrast to the limited funds for addressing the afterlives of militarization in Hunters Point.

The EPA-funded evaluation of the Parcel E-2 landfills did not turn out as residents had hoped. A month after the meeting at the Ruth Williams Theater, the three scientists concluded that the navy's Parcel E-2 remedial action plan, including the landfill cap, met federal environmental standards for exposure risk (although subsequent independent studies have challenged this assessment).[79]

Still, the question of whether the navy's remedy for Parcel E-2 met federal standards is not what Bayview-Hunters Point residents at the meeting had asked. The federal Superfund redevelopment project follows a formal, legal process through which the complex political ecologies of the military shipyard are converted into abstract problems, while professional experts are the only legitimate problem solvers. For many longtime residents, however, the shipyard was not an abstract problem space. Rather, it was filled with memory and politics, entangled with life histories—with family migrations to California and pride in military service, with economic decline in the neighborhood, and with cancers in the family or the lump on one's neck. Rather than a technical-legal analysis, Bayview-Hunters Point residents wanted the landfills evaluated from a different, more expansive framework, one that took into account the ways the military's afterlives intersected with historical racial geographies and their lived experiences in the city. The shipyard was more than a discrete property or stack of risk assessments; it was entangled with personal and collective histories and broader sociopolitical relations. As I have shown in this chapter, Bayview-Hunters Point residents have for decades sought to redefine remediation beyond specific numerical risk levels. Remediation, they have argued, ought to be a reparative project, beginning with an acknowledgment of the harmful environmental and economic impacts of militarization on longtime residents. Residents at the Parcel E-2 meeting in July 2012 also emphasized the neighborhood's history of environmental vulnerability and racial health disparities as analytically significant in the scientists' evaluations of the landfill cap. This understanding of remediation is similar to the way philosopher Ben Almassi theorizes environmental restoration not merely about "ecosystem configurations" but as "attending to victims, perpetrators, and ecological-social relationships in their histories of environmental degradation."[80]

Their critiques and demands, in other words, exceeded the technical-legal boundaries of the federal Superfund program. Indeed, this chapter has pointed to the limitations of the state, or at least of state regulatory agencies, in the work of socioenvironmental repair. No amount of participation, even through the RAB format, was going to deliver the kinds of environmental and economic justice Bayview-Hunters Point residents sought. And yet—and what is, for me, the trickier point—the political strategies and goals adopted by many residents and advocates arguably remained, to a large extent, tethered to state as a provider of justice and reparations, even as their broader visions and desires exceeded what the state could offer. For example, even if the shipyard was fully remediated to residential standards, it would not be free of toxic risk. Remediation would in all cases offer, as Shannon Cram puts it, "a brand of safety haunted by reasonable harm."[81] And, separately but relatedly, creating new jobs for local residents could never address the full economic ramifications of military urbanization in Bayview-Hunters Point. Most broadly, I think, residents were navigating concepts of repair under conditions of environmental and economic irreparability. Still, I think there is much to learn from listening to what residents had to say and how they tried to engage state agencies to do better—how, over the past two decades, people showed up to meeting after meeting, year after year, and continued to articulate the range of harms they believed they, their families, and their community had suffered and some forms of the environmental and social justice they felt they deserved.

4　The Dust of Redevelopment

Angie ushered me over to a chain-link fence, wrapped in green construction tarp, at the edge of the old West Point public housing project.[1] The fence separated us from tall piles of dirt and stacks of lumber—part demolition debris, part building materials for a new, mixed-income housing development called Hunters View. A brisk San Francisco wind whipped around us, carrying a fine dust that caused my eyes to squint. Both old and new housing developments covered the hillside that sloped down from us, stretching out toward the remnants of Pacific Gas & Electric's (PG&E) former power plant and the San Francisco Bay. From certain vantage points I also could see the military's tall gantry crane and new condominiums at the Hunters Point Shipyard.

I came to West Point that day with Marie Harrison to help her present Greenaction's online, participatory environmental justice mapping project to a group of West Point residents. We had driven over from Greenaction's offices, which were, at the time, in the Tenderloin district of San Francisco. After Marie parked her black SUV on a cul-de-sac near the squat, two-story, gray-blue buildings of West Point, she called to her friend, Edna, who came out of one of the buildings. Together the three of us walked up a concrete staircase to a large, hilltop courtyard. A group of women sat on benches to the side of an empty basketball court, and they welcomed us warmly, bantering with Marie and Edna as we filed inside the community room to a cluster of metal folding chairs. Marie took the lead in explaining the mapping project to the group, and I demonstrated how to enter environmental problems on the Identifying Violations Affecting Neighborhoods (IVAN; see note 44 in chapter 3) website or by calling a telephone hotline. Then we opened the floor with a question: What were people concerned about? Natalia, who ran a youth center on the hill, immediately brought up the issue of brown water coming out of the faucets in some older housing units. The environmental problem the group was most worried about that day, however, was the dust: clouds of dust crossing the construction site fence line into their yards; the way the company's sprinkler system, intended to stop the dust's spread, worked on some days but not others; and the construction dust and diesel emissions from large trucks coming to and from the navy shipyard, which was downhill and a mile south of West Point. Natalia shook her head as we entered these complaints onto IVAN's platform. She was in her late forties and, she said, had recently developed asthma, which she connected to the dust of redevelopment.

After our presentation, Angie took me for a walk along the construction fence and asked if I would take pictures of the dirt piles for IVAN's online map of environmental problems. I took several photographs with my phone and later uploaded them to the website. As we walked, Angie explained that construction work starts with a heavy banging at seven or eight in the morning and continues until three in the afternoon. Sometimes the company uses water sprinklers to tamp down the construction dust. In theory, water droplets attach to airborne particulates and pull them to the ground, reducing Angie's and other residents' exposure to dust. After the workers leave, however, the wind continues to pick up construction dust and blow dusty air throughout the yard.

The old West Point units were slated for demolition as part of a larger effort to redevelop public housing across the city, through a program called HOPE SF. Built in 1957 on the foundation of World War II war worker housing, West Point had become "virtually uninhabitable" by the early 2000s due to the lack of maintenance by the SFHA, itself underfunded by HUD.[2] West Point residents lived with mold, cockroaches, appliances that didn't work, and sewage problems, as well as a housing authority that could take months to respond to complaints. To be clear, the new development, called Hunters View, was not simply a replacement for West Point public housing. Rather, Hunters View was a privately managed, mixed-income, mixed-ownership development; when completed, half the units would be for sale at market prices. The new development thus represents what Edward Goetz calls the "neoliberal transformation of public housing."[3]

The website for HOPE SF described the redevelopment program as a "public housing transformation and reparations initiative" and promised one-to-one replacement of demolished units so that, in theory at least, no one would be displaced. Additionally, the construction of Hunters View was scheduled to unfold in multiple phases, to minimize the displacement of West Point households. And West Point residents were also promised relocation housing during construction and assured preferential treatment in moving into the new Hunters View apartments. This seemed reasonable, and indeed some residents had already made the transition. In the car on the way to West Point that day, however, Marie told me about some of the barriers people faced in moving into Hunters View. People could not get into the new apartments without an income or if they were behind on rent, for example, and so Marie knew of West Point residents moving out of Bayview-Hunters Point and into San Francisco's Tenderloin neighborhood, or out of the city entirely. Others, like Angie, preferred to remain in their homes during construction. Why did people stay in West Point, if they were also concerned about construction dust? Some may not have qualified for housing in Hunters View, for the reasons above, and so they stayed on the hill for as long as they could. Yet even if they did qualify, many residents distrusted promises by city agencies, particularly having to do with redevelopment projects. As I have shown, the memory of displacement from urban renewal in the Fillmore remained strong among longtime Bayview-Hunters Point residents, as was a lack of confidence that city agencies had their best interests in mind. Marie spoke to the weight of this memory in an article she published in the *San Francisco Bay View* in 2006, reflecting on her mother's experience with the Redevelopment Agency in the Fillmore. Marie wrote: "My mom, up until the day she passed away, still had her certificate of preference. Like many Fillmore residents, it quietly showed itself to be just another piece of paper."[4] In response to community mobilization in the 1960s, the SFRA had promised displaced Fillmore residents priority in moving back to the neighborhood after urban renewal. Yet for many, as with Marie's mother, this promise remained unfulfilled. These personal stories of displacement through redevelopment alongside memories of the unfinished SFRA-JHC project in Bayview-Hunters Point from the 1960s left some West Point residents hesitant to accept relocation housing—choosing instead to remain in their homes, living alongside the new construction. When I asked an environmental justice activist about this in 2015, they added that people in West Point were hesitant to speak up about construction dust because they were afraid of retaliation—of not being able to move into Hunters View. Indeed, according to a housing rights advocate who served on the HOPE SF Task Force in 2009, "The problem with phased [one-for-one] relocation is that anyone who has a problem with the Housing Authority is getting evicted to make room for relocation on site."[5] West Point residents found themselves stuck between two risky situations—the risk of displacement and the risk of exposure to redevelopment dust—with little faith that government agencies would protect them from either.

As Angie took me along the perimeter of the fence, I noticed a sign announcing the new Hunters View as a "LEED for Homes," project, certified by the U.S. Green Building Council, along with the name of a local company, Bright Green Strategies. Leadership in Energy and Environmental Design (LEED) is a voluntary certificate that rates buildings and land development projects according to various "sustainability indicators," such as access to public transportation or the creation of open space. Indeed, sustainability was central to how city agencies and private actors envisioned the new Hunters View housing development. According to official plans for the project: "Sustainability is one of the core principles for the design of Hunters View, guiding the design of both buildings and site."[6] Did these piles of dirt and construction materials represent the arrival of sustainability on Hunters Point hill? If so, what does it mean that Angie and others still living in the old West Point units experienced the redevelopment project as a health hazard—belonging on a map of environmental injustice?

One of the ways Bayview-Hunters Point residents—and in particular, those living in housing near construction sites along the Hunters Point

Figure 8. Bright Green Strategies' construction fence and signs on the border of West Point and Hunters View, August 17, 2015. Photo by author.

waterfront—have experienced redevelopment is through a seemingly insignificant effect, or by-product: dust. The concerns Marie and I entered onto Greenaction's map that day were not isolated complaints. For the nine years I conducted research for this book—at public meetings, in interviews, at demonstrations, and at community events—I listened to people talk about air quality issues from a number of redevelopment projects throughout the neighborhood, many of them advertised as "green" and "sustainable." What residents experienced as dust is part of a broader category of air pollution, what public health professionals call *particulate matter* (PM). PM is defined as small particles and liquids containing acids, organic chemicals, metals, and soil or dust particles.[7] Exposure to PM, especially fine particles (less than 2.5 micrometers in size), kills. According to geographer Becky Mansfield, in relation to a subset of PM: "In 2018, over eight million people across the world died prematurely due to exposure to particulate matters generated by fossil fuels. . . . Chronic health problems and premature death are attributable especially to fine particles."[8] Ischemic heart disease and lung cancer are among the leading causes of death for Bayview-Hunters Point residents, and residents are hospitalized for asthma and congestive heart failure at much higher rates than elsewhere in the city; all these diseases are linked with long-term exposure to PM.[9] Research shows that even short-term PM exposure can have adverse health consequences.[10]

Most urban studies scholarship does not attend to the ways redevelopment is a material and ecological process that generates its own by-products. Urban studies scholars have focused on the political economy of urban change and, on a more human scale, the lived experiences of gentrification and displacement.[11] Scholars have devoted much less attention to the political *ecologies* of urban restructuring. One exception is anthropologist Catherine Fennell's writing on the demolition of vacant houses in Detroit. In response to severe population decline, in 2014 Detroit embarked on an ambitious plan to "right size" the city, which included tearing down forty thousand old or abandoned buildings. Through this large-scale remaking of the city's mid-century built environment, building materials were reanimated and reintroduced into the local air, including toxic metals such as copper, manganese, iron, and lead.[12] These new respiratory hazards represent how, as geographer Elsa Noterman puts it in her study of redevelopment dust in Philadelphia, "the material remnants of a buried or deteriorating past are reemerging in the reconstruction of cities."[13] Redevelopment in Bayview-Hunters Point involves the demolition of older buildings including West Point and—as I discuss later in the chapter—a large sports stadium, along with the construction of mega-housing developments and new commercial districts. Moreover, environmental remediation (which in many cases must precede new construction) is itself a large, earth-moving operation that has led to a substantial increase in vehicular traffic, and therefore of PM, along the Hunters Point waterfront. A report published in 2007, for example, estimated that 70 percent of diesel emissions in Bayview-Hunters Point that year came from construction activities.[14]

From the perspective of private development companies and city officials, redevelopment dust in Bayview-Hunters Point was a temporary inconvenience on the path to urban improvement. Construction dust issues

that residents worried about were, in general, seen as anomalous, technical issues, which could be fixed with better air quality management plans.
In contrast, many longtime residents experienced and interpreted redevelopment dust as part of a history of racialized environmental vulnerability in the city. Lived experience and family histories had taught many
people not to expect a company's dust mitigation plan to work for them, or
for the city health department to protect them. Recall the political cartoon
in the context of the HPTU rent strike, from chapter 2—the Health Department was closed. In a community that has suffered from health problems linked with industrial overburden and state abandonment for over
half a century, redevelopment dust contributed to and reinforced long-
standing feelings of racialized disposability.[15]

This chapter traces a history of organizing against air pollution in
Bayview-Hunters Point. Between the 1970s and 2000s, residents organized
against the expansion of a sewage treatment facility and against power
plants. Today, environmental justice activism in the neighborhood includes
contesting the dusts of redevelopment. I show how residents have protested
and politicized redevelopment dust, situating dust exposures within the history of neglect and racialized environmental vulnerability in the neighborhood. Environmental cleanup and urban redevelopment are transforming
Bayview-Hunters Point into a desirable destination and place to live, yet,
in the process, residents have found themselves continuing long-standing
struggles to protect themselves from environmental toxicity.

ORGANIZING AGAINST AIR POLLUTION: FROM A SEWAGE
TREATMENT PLANT TO NEW HOUSING

In November 2011 I joined a protest against redevelopment dust outside
the Bay Area Air Quality Management District's (BAAQMD) offices on
Ellis Street, led by Greenaction and the Bayview-Hunters Point Mothers
and Fathers Environmental Justice Committee. A crowd of roughly thirty
people, including residents, Greenaction supporters living in other parts of
the Bay Area, and environmental justice activists from East Palo Alto (another Bay Area neighborhood impacted by industrial pollution, near Silicon
Valley) held signs and chatted on the sidewalk. Eventually Bradley Angel,

the director of Greenaction, gathered the crowd into a semicircle around a microphone stand, while a racially diverse group of Bayview-Hunters Point residents testified to experiences with air pollution in the neighborhood. They detailed their own symptoms and diseases and those they saw in friends and family members, and they connected these illnesses with their urban environment. Redevelopment dust was part of these personal and collective histories of toxicity; it was another layer in the embodied experience of living near so many sources of air pollution over time. Afterward, the crowd chanted, "We want clean air."

Bayview-Hunters Point residents have been protesting air pollution in the neighborhood for decades. In the mid 1970s, neighborhood residents organized to oppose the city's plans to expand the Southeast Treatment Plant, which processed municipal sewage and was located a block west of Third Street, in Bayview-Hunters Point.[16] The expansion project reflected the historical imaginary of southeast San Francisco as a dumping ground: of toxic industries, military waste, minoritized social groups, and in this instance, municipal wastewater. The expansion would increase the amount of sewage passing through Bayview-Hunters Point, from 20 to 80 percent of the city's total, while the plant's diesel generators threatened to worsen air quality for people living near the plant. According to the *Sun-Reporter*, reporting on public hearings on sewage plant expansion in 1975:

> So far city engineers are choosing to ignore community complaints and simply assert that the treatment facility they are proposing won't have adverse effects on the neighborhood. Robert Levy [a city engineer] told Supervisors at last week's hearing that the plant wouldn't smell, wouldn't make noise, wouldn't increase pollution and wouldn't look bad. In support of his last claim he showed a series of slides of facilities in other areas, but the hundred or so people who attended the hearing seemed unimpressed with these aesthetics.[17]

Organized neighborhood opposition was unable to halt the expansion, and by the time I started fieldwork in Bayview-Hunters Point in the 2010s, the sewage treatment plant was releasing the bad smells and air pollution residents had feared. Residents living near the plant—in for some, across the street from it—were often compelled to keep their windows closed because of the smell, and children playing nearby reported headaches.

During periods of heavy rain, the treatment plant released raw sewage into San Francisco Bay, where some local residents liked to fish.[18] In October 2011 I interviewed Aron, the owner of a greenhouse located next to the Sewage Treatment Plant. Beyond rows of trees and houseplants, I could see the circular drums of the treatment plant's "pancake" digesters. I asked Aron how he felt about working so close to the sewage plant. He nodded, "this is an issue," and went on to describe some of the sensory experiences that could be part of his work day. "There is the methane released, which you cannot always smell," he explained, "but periodically there is this burnt coffee smell. I don't know what it is." Aron, who did not live in Bayview-Hunters Point, was less concerned about his own health and more troubled by the long-term impact of the treatment plant on nearby residents. In fact, a 2011 report identified the area around the Sewage Treatment Plant as a toxic *hot spot*, an area of increased cancer risk.[19] The city engineer's assurances in 1975 that the facility wouldn't smell or increase pollution in the area ultimately represented another failed promise to Bayview-Hunters Point residents and workers and another reason to distrust city agencies, including state scientific assessments.

Indeed, although residents in the 1970s did not use the term, by the 2010s the treatment plant was recognized as symbol of environmental injustice. For example, in 2011 I joined a "toxic tour" of Bayview-Hunters Point for local high school students, led by Dr. Tompkins, a longtime environmental and health activist, who had previously served on the RAB. The Southeast Sewage Treatment plant was one of our tour stops.[20] The following year I joined an event for city college students at Heron's Head Park, located across from a power plant (discussed in the following section) undergoing remediation, led by Candace, an Asian American woman who had grown up in Bayview-Hunters Point. Candace spoke about wetland restoration at the park as a form of environmental justice—a small slice of nature replacing industrial pollution in a historically polluted part of town. To make her point, she pointed to the recycling center next to where we stood, where diesel trucks carried away loads of materials every day; then directed our attention to the tall, two-story-high piles of concrete at a nearby concrete mixing plant, just north of the park; and then turned us toward the navy shipyard, to the south, across a small inlet of water. Industrial facilities and hazardous waste sites surrounded us. Afterward, one of the college students raised his hand: "I grew up in Bayview-Hunters Point," he said, "and I was wondering if you could tell us about the smell?" Candace nodded her head and pointed in the direction of the Southeast Treatment Facility. "When I was growing up," she remembered, "I always knew which way the wind was blowing depending on the smell." The treatment plant's emissions were woven into their sensory memories from childhood.

Although they could not stop the treatment plant expansion, Bayview-Hunters Point residents were able to extract concessions from the city in the form of a community center, called the Southeast Community Center. City officials had initially offered to build a park and football field on top of the sewage treatment plant as compensation for the added environmental burdens from the facility. Residents opposed the park, however, on the basis that it would "generate no economic value for the neighborhood" and lobbied for a community center that would provide educational and workforce development programming instead.[21] The Southeast Community Center, located next to the sewage treatment plant at Oakdale and Third Street, was completed in 1988.[22] During the 2010s I attended countless meetings and events in the Alex Pitcher meeting room—a large, high-ceilinged room on the ground floor of the community center, named for an important neighborhood lawyer and civil rights advocate. Today murals adorn the entrance to the building, including portraits of Elouise Westbrook and Espanola Jackson, veteran organizers who had also served on the Bayview-Hunters Point Joint Housing Committee in the 1960s. Inside the Alex Pitcher meeting room, a wall of windows looks north onto a grassy courtyard. Through those tall windows I could often see glimpses of the round digesters of the sewage treatment plant, just past the courtyard—a reminder of the historical wastelanding of southeast San Francisco, even as the community center, with its murals and memorials, represents part of the archive of social movement activism in the neighborhood.[23] This archive is also evident in terms of what's absent in Bayview-Hunters Point: namely, power plants that activists had successfully opposed.

Protesting Power Plants

The struggle against the sewage treatment plant expansion in Bayview-Hunters Point took place at the same time that civil rights leaders and residents of Warren County, North Carolina, laid down in the middle of the highway to prevent trucks from carrying PCBs into their town. During the 1980s the concepts of environmental justice and environmental racism emerged and circulated through news media reports, organizing networks, and public scholarship. By 1994, when San Francisco Energy Company, a subsidiary of a Virginia-based company, applied to the California Energy Commission (CEC) for permission to build a new power plant in Bayview-Hunters Point, the concepts of environmental racism and environmental justice had become part of the lexicon of neighborhood organizers, as had a growing understanding of the connection between industrial emissions and racialized health disparities. Bayview-Hunters Point residents already lived with toxic emissions from an existing power plant, operated by PG&E, located near the shipyard. A second power plant was operating a few miles north, in the Potrero Hill neighborhood.

Residents formed or became active in several local environmental justice organizations, including Bayview-Hunters Point Community Advocates and the Southeast Alliance for Environmental Justice (SAEJ), which worked together to protest San Francisco Energy's proposed plant in Bayview-Hunters Point and later pushed for the closure of the existing PG&E plant. Bradley Angel had already been working in Bayview-Hunters Point through Greenpeace, and in 1998 Angel founded Greenaction for Health and Environmental Justice. Marie Harrison, who had been working with SAEJ to oppose plans for a new power plant in the neighborhood, joined Greenaction's staff that year as a paid organizer. Harrison worked with another resident, Tessie Ester, president of the Huntersview Tenants Association, to organize the Bayview-Hunters Point Mothers Environmental Justice Committee.[24]

Bayview-Hunters Point residents and their allies pursued several strategies to block San Francisco Energy's power plant. Working with Golden Gate University's recently established Environmental Law and Justice Clinic, they developed a toxics profile of the neighborhood and a map depicting the concentration of industrial facilities in southeast San Francisco. Residents also testified at the CEC hearings on the proposed plant in the summer of 1995. At one hearing, Osceola Washington, who had been the first chairperson of the Bayview-Hunters Point CDC in 1965, admonished the CEC: "It's a dump yard out here. This is the dump yard of San Francisco. Everything they don't want, they send it here. . . . They would never build this plant in Pacific Heights or the

Marina District."[25] Washington told the room what everyone already knew: that Bayview-Hunters Point was treated as the city's wasteland. She compared Bayview-Hunters Point with two other wealthy, white neighborhoods in San Francisco to emphasize how the siting of power plants reflected racialized valuations of urban space: "they would never build this plant" in those others places.

At the same hearing, the reverend of a local church testified about the existing PG&E power plant in Bayview-Hunters Point, emphasizing the air pollution people already lived with:

> We have a high rate of cancer, asthma, bronchitis and emphysema in this community. I believe that this is mainly the result of our being continuously exposed to chemicals dumped in the air. Living one-fourth a mile from the PG&E [Pacific Gas & Electric] plant, I hear, see and taste the chemicals, every day.... In the morning the air is so thick with emissions that I can taste it.... My daughter has said to me that it is hard for her to breathe after playing outside. There is a lot of dust blowing around all of the time.... We don't know what chemicals we are being exposed to every day.[26]

Breathing is an act necessary to life; indeed, the breath often symbolizes life itself. Breathing connects us to the world around us—through the breath we take in the world, and release ourselves into the world.[27] The reverend described breathing outside his house as a life-diminishing, rather than life-affirming, activity.

Despite these impassioned testimonials, the CEC continued to support San Francisco Energy's proposal. Still, full approval for the new power plant depended on whether Mayor Willie Brown and the city board of supervisors would agree to lease the land to the energy company. This time, Bayview-Hunters Point residents found allies in city government. Brown had spoken out against the power plant during his recent mayoral campaign. Additionally, in response to the demands of Bayview-Hunters Point residents who were organizing against the power plant and demanding a better understanding of environmental health problems in the neighborhood, the SFDPH conducted a study of cancer rates, which was published a few months after the CEC hearing. Rates of cervical and breast cancer for Black women in Bayview-Hunters Point, the study showed, were double those for women in the Bay Area as a whole.[28] Persuaded by strong local opposition and by the public health study, in 1996 the city board of supervisors sided with Brown and Bayview-Hunters Point residents and voted unanimously against leasing the land to San Francisco Energy.[29] Neighborhood organizing against the power plant had been a success.

Bayview-Hunters Point residents continued to live with emissions from the existing Hunters Point power plant, however. Then, in 1998—in response to new air quality regulations, which would have forced PG&E to pursue expensive retrofits—the company announced it was closing its Hunters Point facility.[30] The Hunters Point plant was set to shut down in 2000 and be remediated by 2002.[31] Yet the closure did not unfold according to plan. Instead, deregulation in California led to an energy crisis and rolling brownouts in 2000. Meanwhile, PG&E kept its Hunters Point plant running to support regional energy needs—in spite of ongoing neighborhood demands to close it down, as had been promised. Then, in 2004, PG&E announced it was filing for another five-year permit to continue operating in Hunters Point, signaling that the plant's closure was nowhere in sight.[32]

Bayview-Hunters Point residents responded with direct action; in December 2004 and again in April 2006, they shut down the gates of the power plant.[33] City officials had been working with PG&E to close the Hunters Point plant, but the April demonstration—organized by Bayview-Hunters Point Community Advocates, Literacy for Environmental Justice, and Greenaction—represented the final push: PG&E agreed to close the Hunters Point power plant the following month. In May 2006 Tessie Ester told a journalist with the *San Francisco Chronicle*, "The smoke has stopped. When I look over at those stacks, and there is nothing coming out, I can't help but cry. We should never have been forced to live like that."[34]

Redevelopment Dust at the Shipyard

A month before PG&E's power plant stopped operating, Lennar began leveling hillsides on the Hunters Point shipyard—on the site known, for remediation purposes, as Parcel A—preparing for new development. Combined, the closure of the power plant and the beginnings of redevelopment at the shipyard marked the transition of the Hunters Point waterfront from an industrial hinterland (a good place for a power plant) to an up-and-coming residential neighborhood. Yet construction work at the shipyard introduced new respiratory concerns for residents living nearby.

The bedrock in southeast San Francisco contains chrysotile asbestos, so municipal regulations require that construction work undertaken there include both dust and asbestos mitigation plans. Undisturbed, asbestos is not harmful to people, but during construction work it becomes airborne and inhalable.[35] Lennar's asbestos mitigation plan included the placement of air monitors along the perimeter, or fence line, of their property. According to the plan, if the air monitors registered concentrations of asbestos above a particular threshold, Lennar's construction work would stop, temporarily, while the company evaluated its air pollution safety measures. However, for the first three and a half months of Lennar's grading work, from the end of April 2006 to the beginning of August of that year, the fence line air monitors did not collect any data on asbestos levels—the monitors simply did not work.[36] When the asbestos air monitors finally did start to collect air quality data, moreover, the data were not analyzed in real time. Instead, an outside lab could take several days to process these data, creating a lag time between days when asbestos levels exceeded state-determined thresholds of safety and work stoppages.[37] Additionally, the air monitors measuring dust levels (which were separate from the asbestos monitors) were not even appropriate for outdoor air sampling. Rather, the company operating the air monitors (which was subcontracted by Lennar) had installed *indoor* air monitors to measure dust on an *outdoor* construction project.[38]

Residents living next to Lennar's new property, along with teachers from the Nation of Islam school located across a chain-link fence from the development site, reported health problems during this time and called for a moratorium on Lennar's construction work until a comprehensive health study could be conducted.[39] The SFDPH responded by issuing three notices of violation to Lennar, for violating its dust and mitigation plans, yet the health department maintained that asbestos and dust levels from construction were safe for residents and schoolchildren. Other city organizations supported the moratorium, however. When I interviewed Dr. Sumchai in 2017, she described a public hearing at a local elementary school, during which teachers and parents testified about kids with bloody noses and asthma. That led to a unanimous vote by the

San Francisco Board of Education on a resolution in support of the moratorium and in favor of an independent health study.[40]

And yet despite residents' health concerns and growing public outcry, it was not until the middle of 2007—when the company's grading work was almost complete—that the city Board of Supervisors held a hearing to address residents' concerns, followed by a health evaluation by the Agency for Toxic Substances and Disease Registry (ATSDR).[41] The ATSDR's report concluded that residents living near the construction site were at a low risk of developing lung cancer from asbestos churned up by Lennar's construction work, reinforcing the SFDPH's assurances that people were not harmed. Yet due to the use of indoor air monitors at the construction site, the ATSDR was unable to interpret reported health issues (the respiratory problems, nausea, and vomiting), which residents believed were connected with the development company's construction dust.

A scientist from SFDPH I interviewed in 2011 was emphatic about the department's position: there was no significant exposure to asbestos beyond background (or ambient) levels during the 2006–2007 events. Toward the end of our conversation, it was clear the scientist found my questions about whether residents were harmed by construction dust misguided. Rather, she believed that some neighborhood activists were stirring up fears of exposure to asbestos as a way to oppose Lennar's redevelopment project. And she advised me to look into other, established causes of respiratory health problems in Bayview-Hunters Point, such as smoking and indoor mold.[42] Indeed, there is no scientific evidence that could connect residents' reported symptoms with construction work on the shipyard (though this is partly due to the faulty and inappropriate use of air monitors). And it is true that air quality issues from construction had become entangled in anti-redevelopment organizing (see chapter 3). Still, I disagreed with the scientist's assessment, which seemed to imply that local residents were dupes, activists manipulative, and environmental toxicity a ploy. To my mind, it could not be this simple. For example, even if residents' relatively short-term exposure to asbestos wasn't concerning to public health professionals, people were still breathing in construction dust during the duration of Lennar's grading work. Moreover, residents were not only upset about the dust. They were also upset that it took state agencies so long to conduct a health study, especially in the context of such egregious air quality violations. And they were egregious: in 2008 the company agreed to a $515,000 settlement with BAAQMD over the developer's failure to monitor and control asbestos dust at the shipyard. Lennar also agreed to update its dust and asbestos mitigation plans and to allow independent third-party air monitors around the development site.[43] This delay was experienced as another instance of the racialized neglect that has characterized the relationship of state agencies with Bayview-Hunters Point residents for decades. When I asked the organizer Bernice about these events in 2011, she told me, "The community continued to complain about headaches, nausea, bloody noses, and dry and itchy skin and eyes. After we protested and protested, we were told by our health department that there were, in their eyes, no long-term effects. Not that we were not affected, but in a word, that it would not last." Bernice emphasized the labor involved in eliciting what she felt was a proper response from state agencies. She also expressed her frustration with what she saw as the dismissal of residents' reported health symptoms. The ATSDR report, and the official response to people's dust complaints, focused on the long-term effects of exposure to *asbestos*, in the form of cancer, but could not comment on other reported health complications, complications that could be linked with breathing in dust and other PM and are often labeled, and ignored, as mere "symptoms."[44]

Still, air quality problems from construction dust at the shipyard continued, which compelled Greenaction and the Bayview-Hunters Point Mothers Environmental Justice Committee to demonstrate at the BAAQMD offices on Ellis Street in November 2011. From the perspective of the protestors at the event, redevelopment dust was not a temporary irritant but part of a cumulative, toxic body burden on neighborhood residents—something people had been struggling against for decades. Marie Harrison and Tessie Ester—both integral to the decade-long struggle against power plants in Hunters Point—were both at the demonstration. In 2003, during the height of the struggle to close PG&E's Hunters Point power plant, Ester had told a journalist from the *San Francisco Chronicle*, "I would like to breath some fresh air."[45] A decade later she was still organizing for better air quality, only this time against the dust from an upscale redevelopment project. Ester, who lived uphill from the old power plant, had brought a group of teenage boys from the neighborhood to the protest. The boys wore surgical masks, simultaneously performing the demand for state protection from redevelopment dust and faulting the state air quality agency for its absence in the neighborhood, since they required personal masks to protect themselves. One of the teenagers held a sign that read, "The air we breathe is toxic. Let us live."

BREATHING ON BAYVIEW HILL

A few years after the demonstration at BAAQMD, residents faced another construction dust issue. Lennar's redevelopment project in Bayview-Hunters Point had expanded to include both the navy shipyard and the adjacent Candlestick Park sports stadium, which would be torn down and replaced with housing, parks, retail space, a performance venue, and office space.[46]

The company's original plan for demolishing Candlestick stadium involved manually taking the stadium apart. In the fall months of 2014, however, residents learned that the San Francisco Planning Commission had quietly approved an addendum to the original environmental impact report (EIR) for the stadium demolition, allowing for the quicker and cheaper method of explosion (the technical term is *controlled implosion*). The explosion was slated to be part of the February 2015 Super Bowl halftime show—a highly visible, nationally televised event. Indeed, a company executive boasted about the stadium explosion: "We'll blow up Candlestick," the executive told a journalist with the online magazine *Vox* in October 2014. "It'll be a good party."[47]

Residents who lived on Bayview Hill, near Candlestick stadium, felt that the planning commission had improperly rushed to approve the EIR addendum, allowing for the stadium explosion without meaningful public deliberation. They saw the explosion as an unacceptable health risk, as the blast would create clouds of demolition dust. And they didn't believe the EIR's assurances that nearby residents would be safe.[48]

Alice, a middle-aged Black woman who lived uphill from Candlestick stadium, learned about the planned explosion at a community meeting in November 2014, after the city planning commission had already approved the EIR addendum.[49] With others from the neighborhood association, Alice walked door-to-door, handing out fliers about the imminent explosion and encouraging people to attend upcoming public meetings to oppose it. The neighborhood association also gathered signatures for an online petition against the explosion and consulted with Golden Gate University's Environmental Law and Justice Clinic, which had a proven record of supporting local environmental justice campaigns. After a heated public meeting at a local elementary school and favorable press coverage of the anti-explosion campaign—and, not insignificantly, as Black Lives Matter protests were spreading around the country—Lennar reverted to its original plan of manual tear down. The Bayview Hill Neighborhood Association's campaign was a success.[50]

In early February 2015, as the manual tear down of Candlestick Park was in its early stages, I met Alice and her neighbor, Jenny, at Jenny's house—a tall, narrow building perched on the edge of Bayview Hill.[51] Jenny was a young mother with a kind smile and chin-length blond hair. She welcomed me into her home and made licorice tea as we waited for Alice to arrive. We sat across from each other on red couches in her living room, seated beneath several windows overlooking the stadium parking lot. Jenny's two young daughters were sleeping down a hallway. I was shocked by how close we were to the stadium and struggled to imagine how the explosion would have affected Jenny and others living in the area.

Alice arrived, dressed smartly in business attire, and sat on a red loveseat next to Jenny. She placed a leather folder on the coffee table. Inside the folder were materials from the anti-explosion campaign, including the fliers Alice had passed out to her neighbors and a printout of the EIR addendum that justified the stadium explosion. "At the meetings, it's called a *controlled implosion*," Alice said, shaking her head, "but you can't control the wind."

An implicit question hung in the space of Jenny's living room, among three sets of raised eyebrows. How could Lennar promise to control demolition dust from this kind of event? Alice opened the EIR addendum from 2014 to an aerial, Google satellite image map of the stadium site. She turned the report around to face me and smoothed the stapled page flat on the coffee table. The map depicted Candlestick stadium surrounded by three concentric circles of different sizes, each drawn in a different color. The circles represented a computer model's predictions of the distance that two kinds of particular matter, PM_{10} and $PM_{2.5}$, would travel under two different explosion scenarios, "still" and "windy" conditions. PM_{10} refers to particulate matter with a diameter of 10 micrometers or less, while $PM_{2.5}$, also called "fine particles" (as previously noted), has a diameter of 2.5 micrometers or less. $PM_{2.5}$ is more dangerous, or toxic, to people because it is smaller and more likely to be inhaled and lodged in the lungs. $PM_{2.5}$ is also lighter in weight than PM_{10} and can travel farther from its point source—which in this case would have been an exploding stadium in an area of the city known for its strong winds.[52]

One aspect of the addendum's map was immediately noticeable. With the exception of a small slice of a residential area called Candlestick Cove, each of the circles stayed within the borders of the stadium's parking lot, and therefore Lennar's property. In other words, the map predicted the explosion blast would be almost fully contained within the stadium site, suggesting that no one who lived near the stadium would be harmed. This seemed unbelievable to me. We were seated so close to the stadium, I thought that if I opened one of Jenny's windows, I might have been able to throw a rock into the stadium's parking lot. Still, the map also showed only three circles, while according to the parameters for the model, there were four possible scenarios (PM_{10} under "still" and "windy" conditions; and $PM_{2.5}$ under "still" and "windy" conditions). As I read later that evening, on the following page the addendum addressed the missing dust circle and introduced a caveat: the model could not predict the radius of $PM_{2.5}$ under windy conditions, the most hazardous explosion scenario. The EIR addendum brushed this caveat aside: "Given the prevailing winds at Candlestick Point which are from the west, the dust would travel over the stadium lot and then out to the bay, where it would disperse."[53]

The map depicted the explosion as an orderly event, with demolition dust either settling within tidy, circular patterns or—as the text of the document assured—spreading out over the ocean. In both cases, according to the EIR addendum, the explosion did not pose a health risk to nearby residents. Alice, Jenny, and I stared at the clean lines on the page together. "There are no people here," Alice said, tapping her finger on the report. "It's like the people who live here don't exist." It seemed as though the map's primary function, similar to the faulty asbestos monitors from 2006, was to keep the company's construction schedule on track.

At the end of our conversation, Alice invited me to join a toxic tour of construction sites in Bayview-Hunters Point the following weekend, organized by the Bayview Hill Neighborhood Association and also led by Malcolm, a longtime activist who had grown up near the shipyard, and who had worked on the stadium explosion campaign. So the following Saturday morning I drove into the Candlestick stadium parking lot and joined a group of about thirty people milling around a small school bus. The tourgoers included staff from multiple city agencies, law students from the Environmental Law and Justice Clinic, and several employees from Lennar. As we loaded onto the bus, Jenny passed out a brochure put together by a San Francisco–based Black health equity nonprofit. The brochure read: "Are You Tired of Seeing the Air You Breathe? Understanding the Environment, Understanding Your Health."

With everyone seated, the bus took off and headed up Bayview Hill. Alice took the microphone and introduced herself. "I'm a resident here, and I want to show you where I live in relation to the stadium." She hoped to convey to Lennar "what our concerns have been this whole time." In particular Alice hoped to establish an understanding, so that when she and others talked about dust, "you have the view of what we're talking about." When we reached the top of the road, Alice pointed to her house and to Jenny's house. The bus lingered on the street, as city staffers and Lennar employees looked politely at the stadium. No one asked questions; the bus was awkwardly silent. After a few moments, the bus wound back down the hill and headed to our next stop: the shipyard.

The route to the shipyard took us around the perimeter of the stadium, while Troy, a senior executive at Lennar, and two other employees from the company handed out their own neatly stapled packets of paper.[54] They had prepared slides on dust control and all the "measures in place," which Troy hoped would address some of Alice's concerns. As Troy spoke, a truck rumbled past the bus, and Alice interrupted him. "This truck coming in and out, is there a washing station?" Troy said yes, there were two washing stations. He elaborated on Lennar's dust control measures, explaining that Lennar also swept the street at least three times a day. Malcolm immediately objected to this image of tidiness, pointing out that he could see dirt on the road outside the bus windows right then. Alice nodded, adding that the sides of the road were muddy. She explained: when the roads aren't cleaned up, the mud dries out and turns to dust. "That's what gets airborne when you guys are off site after hours, and on the weekend," she said to the employees from Lennar. Troy responded by reiterating that Lennar has "measures in place." He also told Alice she was pointing to a city-owned lot, which was not Lennar's responsibility.

At this point, a woman seated toward the rear of bus, wearing a neon yellow vest, spoke up and introduced herself as a staffer from BAAQMD. She said that BAAQMD had been coming "out here" for years. She pointed out that businesses in this area of the city often fail to water down their dust, and trucks often idle along the road, releasing diesel emissions. BAAQMD doesn't monitor these businesses, she explained, because they're "too small"—they have too few employees. A staffer from the Department of Public Works, seated across the aisle from the woman, added that some of the streets in this part of Bayview-Hunters Point have never been paved. The city agency staffers appeared to respect Alice and Malcolm's concerns about dust. The BAAQMD employee, for example, carried a clipboard and made notes, which she presumably took back to her office, perhaps discussing them with a colleague or supervisor. At the same time, the city staffers also confirmed Troy's point that Lennar was not liable for all of the dust particles and PM Alice and Malcolm were concerned about.

What struck me later, in reflecting on this exchange, is that everyone on the bus was looking at the same dusty truck, the same dusty street. And yet the meaning of what people saw varied widely. For Alice and Malcolm, the dust clouds churned up by Lennar's truck represented clear and incontrovertible evidence of a community at risk. In contrast, Troy saw the relative absence of dust and the workings of "measures in place." He also pointed to the limits of Lennar's responsibility, since the company was not liable for dust produced by other businesses. It was never entirely clear to me what the state agency staffers thought of the tour, but I have seen other state employees draw sharp lines around their obligation to Bayview-Hunters Point residents. At an environmental justice task force meeting in August 2015, for example, a staffer from the California Air Resources Board responded to complaints about redevelopment dust at Lennar's construction site on the shipyard. "The Air Board doesn't work on the weekends," the staffer said. "They can't just send someone out if a complaint is filed on the weekend." Marie Harrison was sitting to the side of the room, leaning against a wall. Without missing a beat, she replied: "You are aware that this is the Bayview and the wind doesn't stop on the weekends?" Marie's sarcasm highlighted what she saw as the absurdity in how the Air Board staffer delimited the agency's regulations as only in effect between Monday and Friday. Her comment pointed to the ways dust accumulates in the gaps and absences, if not the outright refusals, of state regulation in Bayview-Hunters Point.[55]

Toxic tours like the one Alice and Malcolm were leading tend to enact a common refrain in environmental justice campaigns: come here and experience, with your own senses, what needs to change. Smell the air. Drink the water. See how close we live to this factory, this treatment plant. Marie made a similar claim at the 2012 meeting to discuss Parcel E-2 at the Ruth Williams Memorial Theater. "Instead of relying on the Navy's science," she had told a panel of professional experts, "come talk to someone out here, who has watched people suffer and die." Sociologist Phaedra Pezzullo writes, "Through the rhetorical performance of a toxic tour, people, places, processes, and things may seem more tangible to us and, thus, we may be more persuaded to identify with or believe in their existence, significance, and their consequence." For Pezzullo, "being present, as a mode of advocacy, suggests that the materiality of a place promises the opportunity to shape perceptions, bodies, and lives."[56] The exchange between Alice, Malcolm, Troy, and the woman from BAAQMD about dust on the side of the road, however, demonstrates that being in a place, and indeed the materiality of a place, is not always evident, let alone persuasive. Rather, what counts as evidence of harm is an effect of what Murphy calls "regimes of perceptibility," or the ways a "discipline or epistemological tradition perceives and does not perceive the world."[57] The map of dust scenarios in the EIR addendum that Alice showed me in Jenny's living room relied on a technology of quantitative computer modeling—a regime of perceptibility—that rendered potential dusts from the stadium explosion legible only within discrete, circular areas, neatly contained within Lennar's property lines. The map was part of a broader assemblage of knowledge practices, technologies, and political and economic investments that produced, or aimed to produce, the stadium explosion as a harmless event—indeed, an event that could be made into a global spectacle (the Super Bowl half-time show) and referred to, openly, as a "party." Over tea at Jenny's house, on the other hand, Alice had interpreted the near-miss of an exploding stadium within a systemic pattern of racialized environmental vulnerability in the neighborhood. "This," she said, pointing to the map of PM circles, "is just another assault on the community." Alice's use of the phrase "just another" communicated how she saw the explosion as part of a larger, systemic pattern, while the word "assault" indicated that she perceived construction dust as a form of violence.

Alice, Malcolm, and others involved in the campaign against the stadium explosion also connected residents' exposure to dust with the emerging Movement for Black Lives. In August 2014, a month before the city planning commission approved the EIR addendum, a white police officer killed eighteen-year old Michael Brown in Ferguson, Missouri. Demonstrations erupted across Ferguson after Brown's death and again four months later, after a St. Louis County Grand Jury decided not to indict the officer who had killed Brown. The protests were part of a wave of activism against anti-Black police brutality that summer and fall, which included the killing, in July, of Eric Garner, who died in a police chokehold while repeating "I can't breathe" eleven times, and the shooting of Tamir Rice, a twelve-year-old boy, in November. The hashtag #BlackLivesMatter had circulated on Twitter since at least 2013, but by the fall of 2014—when the city planning commission approved the stadium explosion—"Black Lives Matter" was a household phrase.[58] Bayview-Hunters Point residents and advocates who showed up at meetings on the stadium explosion in 2014 articulated their demands within the lexicon of this new movement, situating their vulnerability to the dust of redevelopment as part of a larger pattern of racialized state violence. As a resident told a television reporter in January 2015, the proposed explosion was a "Black Lives Matter situation."[59] When I spoke with Malcolm over the phone about the stadium explosion, a few weeks before the toxic tour he led with Alice, he told me, "We need to breathe also." "We Can't Breathe," a collective remaking of Garner's last words, had by then become a refrain in social media posts and at protests, emphasizing the pervasiveness of what Christina Sharpe calls "the climate" of anti-Blackness.[60] Now Malcolm linked it to the stadium explosion and to racial-spatial inequalities producing social and environmental vulnerabilities in Bayview-Hunters Point more generally, explaining, "It's the same thing in a different form—institutional racism, gentrification, unemployment."

Dust clouds, dusty trucks, the near-miss of an exploding stadium, and a perceived ambivalence from city agencies reinforced, for many residents, a long-standing feeling of disposability. Their lives were being put at risk, they sensed, for redevelopment projects which were also experienced as risky in another sense—many residents were unclear how these projects would improve their lives, let alone whether they would be displaced by them. Moreover, in an additional twist, many of these urban development projects were advertised and discussed through discourses of greening and sustainability.

REBRANDING REDEVELOPMENT

In her study of the origins of urban greening, sociologist Hillary Angelo details how urban greening projects came to represent "moral goods," which are "beneficial to all and to all in the same way." Instead, as she shows in the case of Germany's Ruhr Valley, urban greening projects have historically functioned as technologies of social control that aimed to produce particular "class-specific ideals about what constitutes good cities and citizens."[61] The social imaginary of nature as an apolitical and universally beneficial good thus "forecloses discussion of [urban greening] projects *as social projects* and, therefore, of their possible benefits or negative effects."[62]

In the years I spent researching this book, I saw how state agencies and private companies increasingly turned to discourses of nature, highlighting greening and sustainability as central components of remediation and redevelopment in Bayview-Hunters Point. In all these cases, nature worked to represent state agencies and companies as well-intentioned providers of public goods, rather than engaged in the work of urban transformation with fundamentally social, and in some cases unequal, implications.[63] The navy, for example, often presented itself as an environmental steward, describing remediation as a story of ecological renewal rather than a project of risk management and property transfer.[64] I encountered this on a bus tour of the shipyard in 2011, led by the navy's BRAC coordinator at the time. As the bus meandered through the shipyard, the coordinator described the history of particular buildings and explained the military's ongoing remediation efforts. Our final stop was a view of Parcel E-2. As we idled by a chain-link fence surrounding a weedy field, marked by yellow signs with the symbol of radioactivity, the coordinator explained that Parcel E-2 was "slated to be safe for open space" and would include a new wetland habitat. "With cleanup," he said, "you will have a beautiful, healthy ecosystem." This was a complicated statement, to say the least, since (as discussed in the previous chapter), the navy's remedial actions for Parcel E-2 involve leaving toxic waste dumps, with radioactive waste, onsite. It was also not an isolated comment. Later that year I watched a navy engineer give a presentation at the Bayview-Hunters Point YMCA that advanced a similar narrative. During her discussion of Parcel E-2, the engineer proudly announced that the navy had removed forty thousand cubic yards of contaminated soil from the area around the parcel's landfills, and went on to speak enthusiastically about recent sightings of coyotes and geese on the base. Her message seemed to be that "nature" had returned to the shipyard as a result of the navy's remediation work. A glossy fact sheet on Parcel E-2 handed out at a remediation meeting in 2017 attempted the same rhetorical work: beneath a photograph of an avocet—a bird typically found on lakeshores and tidal flats—a caption read: "The cleanup solution at Parcel E-2 will allow native species, like the avocet, to thrive at the shipyard." This discourse of nature did not just come from the navy, either. A scientist who worked for a state agency monitoring the navy's work in Hunters Point also emphasized similar ideas of nature when I interviewed him about Parcel E-2 in 2011: "At the end of the day," he told me, "we're going to get the property to a point where people can use and enjoy it, even wildlife and critters will want to use this property. We will have a really high quality habitat for the critters." And in a similar vein, the San Francisco Office of Community Investment and Infrastructure (SFOCII; the successor agency to the SFRA), in its "sustainability plan" for the Hunters Point Shipyard, labels the future Parcel E-2 a "grasslands ecology park."[65] Such framings seek to associate hazardous waste with the shipyard's past while eliminating it from ideas or projections of the military base's remediated, redeveloped future—having become a site of nature, the conceptual

opposite of pollution—welcoming to native wildlife or "critters." I was reminded, on all these occasions, of Shiloh Krupar's argument that "nature
serves to contain the residues of war."[66] It is unclear whether the intended
audience for these pronouncements included longtime neighborhood residents, although from what I observed, residents who attended navy remediation meetings ignored such references to nature. After the navy engineer's
presentation on Parcel E-2 in 2011, for example, no one from the audience
asked follow-up questions about geese or coyotes. Rather, a Black man
seated near the front of the room wanted to know, "How much radioactive
waste was taken off side, and how much has stayed on site?" He followed up
with the question, "Why don't you dig it all up?"

This dynamic is even more pernicious when greening is instrumentalized by private, for-profit development companies. In the years that
I worked on this book, urban greening and sustainability were increasingly used as branding strategies, and they functioned to obscure or at the
very least soften the profit motives at the center of corporate interest in
Bayview-Hunters Point real estate. In some cases, as with the previous examples, ideas of nature were used to represent new development as a de-
finitive break with the area's industrial, military past; relatedly, corporate
uses of nature could also be seen as a way of reducing political controversy
in order to attract global financial investment and sell homes to regional
homebuyers, who desire nature woven into the urban experience.[67] In
2008, for example, green became the central color on Lennar's website for
its development project at the shipyard. Web pages featured computer renderings of tree-lined streets, people walking and cycling, and text advertising eco-friendly design features. This new green branding strategy was
implemented the same year the company paid a hefty fine to BAAQMD
for air quality violations during its grading operations of 2006–2007. Additionally, the financial and housing foreclosure crises, beginning in 2007,
had significantly decreased the value of Lennar's holdings and led to an
industry-wide plummet in new home construction.[68] The company's new,
green branding campaign could be understood, as Krupar puts it in her
study of the uses of nature in the remediation of Rocky Mountain Arsenal
in Colorado, as an "image management strategy."

In 2012 Lennar was certified as LEED-certified "Neighborhood Design" by the U.S. Green Building Council. Ideas and images of nature and

sustainability became even more pronounced on the company's website three years later, in 2015. Aerial renderings again depicted the future development project as a large swath of green, now advertising "spirited neighborhoods," "arts and innovation," "open space," and "sustainable living."[69] These advertising strategies exemplify what Miriam Greenberg calls "sustainability branding," a phrase she uses to describe how businesses and entrepreneurial city governments rely on sustainability discourses and practices to gain market advantage, or a "sustainability edge."[70] Sustainability branding, Greenberg writes, uses specific images, keywords, sustainability metrics, and green building certifications (such as those offered by the U.S. Green Building Council) as forms of market differentiation, often obscuring "the real complexity of cities' socio-ecological entanglements" and "rendering other, non-market-oriented sustainability goals less attainable."[71] Indeed, the 2015 makeover of the company's website was introduced the same year Lennar raised money for both its Hunters Point and Treasure Island development projects through a program offering EB-5 visas to foreign investors. Through the U.S. EB-5 Immigrant Investor Program, foreign investors gain U.S. visas in exchange for significant capital investments in the United States, in this case, at the Hunters Point Shipyard.[72] Computer renderings of the military base as a green, sustainable urban development project would have helped investors see the military base and Superfund site as a good investment.[73]

Lennar's rebranding strategies did more than obscure the shipyard's complicated ecologies; these strategies have also worked to dissociate the shipyard from its historical relationship with Hunters Point. (As previously discussed, Hunters Point was first developed as a residential neighborhood for shipyard workers during World War II, and until the 1970s, the shipyard supported the local economy.) In 2014 the company changed the name of its development project from The Hunters Point Shipyard to The SF Shipyard[SM]. According to the president of The Mark Company, a San Francisco–based firm hired by Lennar to create a new branding strategy for the shipyard, "part of the branding has been doing away with the 'Hunters Point' part of the name, a place that many Bay Area residents associate with the poorly built public housing that is being rebuilt as part of the shipyard redevelopment."[74] As noted in chapter 1, because of the SFHA's segregationist policies, historically public housing in Hunters Point was filled with Black residents. Renaming the shipyard development project effectively marked the shipyard as *not* Hunters Point, through a branding strategy that arguably relied on the historical relationship between Blackness, risk, and property value.[75] Writing about the history of the U.S. housing market, Keeanga-Yamahtta Taylor notes that "the intensely subjective process of determining the value of property or a neighborhood [. . .] was inherently informed by the presence or absence of African Americans."[76] The company's new branding strategy could be understood as reflecting and reinforcing the racial logics of property value described by Taylor, through which value has historically been linked to social and physical distance from Black people and Black spaces.[77]

On a Saturday afternoon in March 2016, I went to the main homebuying office for Five Point, a Lennar affiliate company, on Parcel A, interested in the ways the company presented the military base and Superfund site to potential homebuyers. A mildly distracted sales representative took me on a tour of three different, attractively staged units. At the end of the tour, the sales representative handed me glossy promotional material on thick cardstock. One of the promotional handouts was titled "Surveillance." On the handout, against a dark background, the company described The SF Shipyard[SM] as a "Smart-City" community, with "an intelligent surveillance program." The surveillance system included private video cameras and, at least for the time being, policing of the development by a private security force.

This was, it seemed to me, a form of what Simone Browne calls "racializing surveillance" or "a technology of social control where surveillance practices, policies, and performances concern the production of norms pertaining to race and exercise a 'power to define what is in or out of place.'"[78] Whereas ideas of nature communicated that the military base was environmentally safe, surveillance technologies would have assured potential homebuyers that the development was safe from Hunters Point. Meanwhile, some residents were continuing to protest the health impacts of other development projects.

"IT'S LIKE YOU ARE TELLING ME MY LIFE DOESN'T COUNT"

The Bayview-Hunters Point Environmental Justice Task Force meeting in August 2018 was delayed as a Greenaction staffer struggled to get the audiovisual equipment in the Southeast Community Center to work. Organizers had already rearranged the metal folding chairs in a large circle, with overflow chairs radiating from the main seating area, as they did for all the task force's monthly meetings. They wanted to promote dialogue, giving people the chance to speak to one another directly. That day's meeting was well attended, and most of the chairs were full. The meeting agenda centered on recent revelations of falsified radioactive soil samples at the shipyard. However, the first agenda item concerned a mixed housing and commercial development project on an area of the waterfront just north of the shipyard, called India Basin. Because of their experiences with construction dust at other sites along the waterfront, Bayview-Hunters Point residents were concerned about impacts on local air quality from the redevelopment work. Two representatives from the development company, BUILD—a young white woman and an older white man—had come to the meeting to convince residents in the audience of the benefits of the project. They were accompanied by another white woman from SFOCII. The woman from BUILD anxiously tried to help the Greenaction organizer fix the audiovisual system, but to no avail; she would have to talk without her accompanying slides.

Somewhat flustered without images to refer to, she launched into a description of the development's amenities: 1,575 units of housing, 25 percent of which would be classified as affordable; 200,000 square feet of commercial space; 14 acres of parks and open space; and a public market, which she compared to Pike's Place—a Seattle waterfront attraction. Seated near her, the man chimed in with additional details. He held a printout of the slides and gestured at it frequently, indicating that the visual aesthetics of the project were important to understanding these benefits.

The first question from the audience, after the presentation, came from a resident—a middle-aged Black woman wearing a Greenaction T-shirt and taking notes—who wanted to know where, exactly, the proposed development was located. The woman from BUILD skimmed through her notes and glanced at her colleague for help, while the resident in the Greenaction T-shirt suggested, kindly though unsuccessfully, a street intersection and a landmark, as if she could jog the younger woman's memory. Eventually the man from BUILD, after flipping through his notes, proffered the four boundaries of the property parcel. It was strikingly clear, in this exchange, that the BUILD representatives knew Bayview-Hunters Point not from lived or embodied experience, but through computer images and the bird's-eye view of a planner's map, or what philosopher Henri Lefebvre calls "representations of space."[79] Representations of space produce abstract, empty, interchangeable, alienable space. For Lefebvre, "it is the dominant space in any society" and is, as geographer Andrew Merrifield elaborates, "tied to the relations of production and the 'order' which those relations impose.'"[80] Within capitalist relations of production, representations of space produce and exhibit economic, class power. Still, the abstract space of planners and engineers can be understood as dominant in the way social theorist Antonio Gramsci defines hegemony as a form of dominance—abstract space is a continual achievement, requiring work, and is open to contestation and change, rather than being total or complete. As the discussion continued, it became clear that Bayview-Hunters Point residents at the meeting rejected BUILD's representations of space; indeed, they were unimpressed by the whole presentation. Another Black woman at the meeting, for example, questioned BUILD's definition of affordable housing. She pointed out that the affordable options in BUILD's development project were one-bedroom apartments, which are not suitable for families, adding that most residents who live near the development site cannot even afford what qualifies as "affordable housing" in San Francisco. Yet another woman raised questions about sea level rise and the impact of climate change on BUILD's waterfront property, telling the presenters, "I don't want your pictures, I want to know what kind of engineering you're doing."

Still, most of the discussion following BUILD's presentation focused on redevelopment dust. Sheridan, an organizer with Greenaction, stood up and read a passage from the draft EIR for BUILD's project. Sheridan spoke deliberately, slowing down to enunciate a particular sentence in the report that admits the development will have "significant and unavoidable impacts" on local air quality. Sheridan reminded the developers that many people who live near the development site already suffered from asthma and other respiratory diseases, and that air pollution from new construction would likely compound these existing health issues. The man from BUILD nodded after Sheridan spoke. "Yes," he said, "these are unavoidable impacts. Some impacts are unmitigatable. But we think the benefits outweigh the environmental impact." During this exchange, Marie Harrison was sitting to one side of the room, her hands resting gently on her portable oxygen tank. She responded sharply to the man from BUILD. "Did I hear you say that the benefits of housing and parks outweighs people's health? Our lives? It's like you are telling me my life doesn't count." During the early 2000s, Marie had been part of the group of demonstrators who, in the cold San Francisco rain, had chained themselves to the gates of the PG&E power plant to protest of its ongoing operations. Here, she spoke to a continued experience of disposability, only in the context of what was being described as a visually attractive, eco-friendly development project. A few minutes later, Angela, another Black resident and the director of a neighborhood-based advocacy group, elaborated on Marie's comments.[81] "This is an environmental justice meeting," she said to BUILD's representatives. "I want you to hear the story you are telling us, how out of touch you are. How offensive and privileged you sound. You are cleaning up that land for someone else. The people living here all these years, they will suffer. Only when you can bring in people who don't look like me, then you can clean it up." Angela pointedly called attention to BUILD's project as a form of gentrifying displacement—not a universally beneficial project, as the company's employees were suggesting. She also emphasized the political ecologies of urban improvement and the ways redevelopment can produce racialized environmental harm.

During the meeting, Leaotis, a longtime Black resident and Greenaction organizer, stood near the microphone at the front of the room, arms folded. Now he offered his own take. "I've been here since '66," he reminisced. "I used to play down there. Dirty mountains. We were six, seven, eight years old." Leaotis was remembering BUILD's development site when it was an informal dumpsite, used by kids living in housing projects near the shipyard as a playground. In contrast to the developers from BUILD, Leaotis knew the Hunters Point waterfront intimately. The company's development site was part of his biography and childhood memories; it was what Merrifield, writing about Lefebvre's theory of space, calls lived, every day, "alive" space, with an "affective kernel or centre."[82] Leaotis's memories served as a reminder of the historical abandonment of Bayview-Hunters Point and the lack of proper playgrounds for kids, even as he pointed to lived experiences that exceed narratives of environmental suffering. That is, Leaotis was also describing the pleasures of childhood and what geographer Tianna Bruno has called "living in the meantime of repair."[83] Still, this was an environmental justice meeting, and Leaotis continued, "I see your pictures, but where are we? Our community. We're getting shoved out by that housing." He was referencing BUILD's visuals—implying he had seen them previously—and called out the development company for both discursively ("where are we" in your images?) and materially ("we're getting shoved out") displacing the longtime residents. Leaotis's comments reinforced the collective assertion from other residents in the room that afternoon: they did not see how BUILD's project would benefit them, yet they felt that the company and the city (in the person of the woman from SFOCII, who sat quietly) were sacrificing them to another upscale housing development.

I wondered about the pictures Leaotis had mentioned, and so later that evening I searched online for BUILD's India Basin website. The website depicted the development project as a small-scale ecotopia: condominiums with floor-to-ceiling glass walls faced a natural-looking shoreline that seamlessly transitioned from attractive, light-filled housing to a tidal marshland. These images were filled with people walking dogs on trails, kayaking in the Bay, and seated at open-air cafés. When I looked more closely, I saw what Leaotis was talking about. At best, one could say that the company presented an image of racial diversity, but most of the people in its mockup landscape designs were white. Still, to the extent that the website included Black people in its designs, even as the company's representatives appeared unconcerned about the environmental health issues raised at the meeting that day, Brandi Summers's concept of *black aesthetic emplacement* is apt, particularly the ways "blackness accrues a value that is not necessarily extended to Black bodies."[84]

The political ecology of urban redevelopment includes the ways particulate matter becomes part of the bodies and anxieties of people living near redevelopment sites.[85] In Bayview-Hunters Point, dust and other PM were generated only for the duration of specific construction activities; in contrast to the decades-long emissions from power plants or the sewage treatment plant, these were what public health professionals would consider short-term exposures. Still, for many residents, they were not separate from the broader, racialized accumulation of toxicity in southeast San

Francisco, perpetuated, in part, through uneven enforcement of environmental regulations and what anthropologist Chloe Ahmann, in a different context, calls a "widely disaggregated regulatory sphere whose partitions have for years served corporate power."[86]

The concerns with redevelopment dust over the past twenty years in Bayview-Hunters Point represent a transition in environmental justice organizing in the neighborhood, from fighting against toxic facilities such as sewage treatment facilities and power plants to protesting the health impacts of upscale, often "green" urban developments. These new environmental health concerns emerged within the context of the revaluation of southeast San Francisco—from a waste-able space to a string of gentrifying neighborhoods. From the perspective of the Bayview-Hunters Point residents who organized environmental justice campaigns and who showed up at meetings and protests, this revaluation of space had not translated into the revaluation of their lives; rather, they felt their concerns were ignored in favor of corporate development projects. Their experiences indicate how environmental cleanup and green redevelopment can reproduce racial and spatial inequalities, and can do so by producing new environmental vulnerabilities.

The environmental amenities, green certifications, and natural landscaping of The SF Shipyard[SM] and BUILD's India Basin project exemplify Greenberg's notion of market-oriented sustainability: "a strategic branding device more than an ideal," a form of sustainability "instrumentalized to sustain the competitive environmental for capital."[87] Residents' exposures to the dusts of redevelopment were not a direct consequence of market-oriented sustainability; rather, these exposures stemmed from the political ecologies of large-scale remediation and redevelopment activities more generally, combined with the uneven enforcement of state environmental and health regulations, along with a more pernicious, underlying assumption of the disposability of residents living near redevelopment sites. Arguably, this undercurrent of disposability allowed state agencies and private companies to treat redevelopment dust as a temporary inconvenience—a minor trade-off relative to the benefits of postindustrial, postmilitary redevelopment and the economically productive, nature-oriented reuse of a contaminated urban waterfront. Yet many Bayview-Hunters Point residents experienced redevelopment dust as more than a minor health irritant or a setback on the path to urban improvement. Rather, redevelopment dust represented yet another burden for people living in a historically segregated and industrially overburdened neighborhood. Even if redevelopment dust exposures were ultimately temporary—lasting only for the duration of each construction project—many residents saw those dusts as evidence of the continued disposability of Black life in San Francisco, now in the context of gentrification.

Still, the fact that The SF Shipyard℠ and BUILD's India Basin project incorporated signifiers of nature and sustainability as central to their development designs and branding strategies matters. Whether government-led or corporate, greening and sustainability represent a dominant modality in the redevelopment of Bayview-Hunters Point today. Green discourses and practices are not separate from the inequalities and vulnerabilities produced by urban redevelopment in Bayview-Hunters Point; rather, they are part of the ways the asymmetrical power relations undergirding redevelopment, and produced by it, are obscured. The widespread imaginary of nature as an unquestioned public good also puts some residents in a difficult position of opposing or at least expressing complicated feelings about projects that might otherwise appear as progressive urban improvements.

Over the past twenty years, some residents have worked to visibilize and protest the dust and PM generated by and distributed through urban redevelopment. Their political campaigns and critiques offer important reinterpretations of toxic cleanup and redevelopment in southeast San Francisco today. Namely, they have insisted that cleanup and redevelopment are not simple stories of progress and urban improvement. Rather, these projects, along with the unevenness of state environmental regulation, represent another chapter in the neighborhood's history of racialized toxicity.

Conclusion

REPARATIVE ENVIRONMENTAL JUSTICE

In 2021 the San Francisco Recreation and Parks Department began the process of cleaning up and redeveloping a small, contaminated waterfront property located in between The SF Shipyard℠ and BUILD's India Basin housing developments. For decades this former shipbuilding site, located at 900 Innes Street, was fenced off and unused. Municipal interest in acquiring and remaking the site as a wetland and a public park was connected to regional wetland restoration efforts. And in an area of the city with few green spaces, the social equity dimension of the new park cannot be overlooked. Still, it is hard to separate the city's attention to this greening project from the larger processes of postindustrial redevelopment and green gentrification taking place in southeast San Francisco and Bayview-Hunters Point. In a testament to the significance of the southeast waterfront for San Francisco's future, the park, budgeted for $140 million, will be "the most expensive park in city history."[1] Moreover, it is impossible to untangle the construction of the park from the large residential real estate projects on either side of it. It is very likely the new park will increase real estate values for development companies and for the homeowners who can afford to live in the new housing.

A different kind of urban greening took place on a hillside park not far from 900 Innes Street. On a foggy Saturday morning in 2012, I joined a middle school girls' gardening program at Adam Rogers Park, located near the top of Hunters Point hill. The park is named for an influential Black organizer who, during the 1966 Uprising, worked with Hunters Point youth to deescalate the potential for violent clashes with the police.[2] June, a Black woman from Oakland who was the manager of the girls' gardening program at the time, welcomed me at the garden's gate and showed me around, before I helped the girls transplant seedlings into pots for an upcoming plant sale.[3] There were raised garden beds, a few work sheds, and a gently sloping hillside lined with fruit trees and flowers.

June had paused in her tour of the garden at a grassy area near the raised beds and described her vision of building a wooden platform in the shape of Africa—a place the girls could use for meetings or as a stage for events. By then the fog had cleared, and I saw that the garden looked down on the shipyard. During the second half of the twentieth century, the military base was a node in the transnational circulation of people and things, shaping labor migrations and urban development, processing radioactive ships and the fallout of nuclear weapons, building up and then unraveling the neighborhood economy. Bayview-Hunters Point, and particularly the windy streets along Hunters Point hill, is intimately tied to these historical processes and geographies of militarization. Yet June's idea for the platform reimagined the park on the hill within a different transnational spatiality—the platform in the shape of Africa told what McKittrick calls a "different geographic story" of diasporic connections, representing a geographically expansive Black sense of place in this once-neglected corner of San Francisco.[4]

The gardening program at Adam Rogers Park represents a small but meaningful example of the neighborhood-based movements for environmental repair, urban development, and racial justice this book has traced.[5] In the 1960s Bayview-Hunters Point residents responded to and opposed the state violence of police killings, substandard housing, and the bulldozers and eviction notices that came with urban renewal. They developed organizations and programs to create decent, affordable housing and a healthier, safer living environment, by and for Black San Franciscans—part of a larger social practice I called Black counterplanning. Community control and Black self-determination were central aspects of their counterplans and visions of urban development, even as organizers often worked through state-funded programs or established strategic alliances with state agencies. In subsequent decades, Bayview-Hunters Point residents continued organizing for a healthy living space—opposing, and ultimately demanding mitigations for, the expansion of the Southeast Sewage Treatment Plant in the late 1970s, and organizing against power plants in the 1990s and 2000s. Today, in challenging the navy's remediation work at the shipyard and protesting the dust generated by redevelopment projects on the waterfront, residents have shown how the massive influx of private and state investment in southeast San Francisco has had a complicated impact on longtime Black residents, and that these spatial transformations carry with them a number of uncertainties about what the future of the neighborhood will look like and the place of Black San Franciscans within it.[6] The gardening program at Adam Rogers Park is just one of many grassroots efforts and programs that stake a claim in that future.

The book has traced a history of racial uneven development in southeast San Francisco, and the ways Black residents of Bayview-Hunters Point have organized around urban environmental justice, broadly defined. I have shown how residents worked with and in opposition to state agencies to make Bayview-Hunters Point a better place to live. Theirs is an ongoing struggle, one currently being redefined by a new generation of activists and allies.

One emerging framework residents have begun to engage with is the case for reparations and its relationship with long-standing environmental justice concerns. For example, in February 2022 a few dozen demonstrators gathered on a busy street corner in Bayview-Hunters Point to protest, once again, the navy's cleanup project at the shipyard. The demonstrators demanded the excavation and removal of all radioactive waste from the shipyard. They also demanded community oversight of the remediation process—precisely what was lost when the navy disbanded the RAB in 2009. Yet the demonstration was about more than military waste. According to Kamillah Ealom—who was raised in Bayview-Hunters Point and organizes with Greenaction for Health and Environmental Justice—these demands "are a form of reparations because we are dying out here. We're begging for clean air as reparations at this point. As well as quality living, housing, and employment. The basic needs that people right on the other side of the City don't have to worry about."[7] While reparations are often thought of as a check, or as a specific monetary amount, Ealom defined reparations as establishing some of the basic conditions for a flourishing life: clean air, decent housing, and economic security. What it might take to establish those conditions, however, likely requires an entire reimagining of U.S. society.

ENVIRONMENTAL JUSTICE AS REPARATIONS?

Between 2021 and 2023, as I was finalizing this book, the California Reparations Task Force held public meetings (first on zoom, and later in person) to gather public comments and personal testimony and hear scholars, activists, and lawyers present on the myriad injuries of slavery and its legacies in California. The task force was established through state legislation, AB3121, adopted in September 2020, following the summer uprisings in response to the murder of George Floyd. The task force was charged with "studying the institution of slavery and its lingering negative effects on living African Americans" in the state and recommending "appropriate remedies of compensation, rehabilitation, and restitution."[8] In June 2023 the task force released its final 1,080-page report, which—echoing Ealom's demands—includes a chapter on environmental pollution, inadequate public infrastructure (such as access to clean water), and climate change.[9]

The California report is the most detailed state-level study on reparations, but it joins a long history of reparations claims and debates in the United States, which historian Robin D. G. Kelley discusses in his book *Freedom Dreams*. Kelley shows how "dreams of reparations" throughout U.S. history have been "never entirely, or even primarily about money," but rather "part of a broad strategy to radically transform society."[10] That is, reparations have included individual payments, but also projects of building autonomous Black institutions, improving community life, and eliminating institutional racism. Most recently, in 2019 the Movement for Black Lives published its platform on reparations, the *Reparations Now Toolkit*. In its accounting of the breadth of what Saidiya Hartman calls "the afterlives of slavery," the report includes the need for reparations for "the wealth extracted from our communities through environmental racism, food apartheid, housing discrimination, and racialized capitalism."[11] *Reparations Now* also references journalist Ta-Nehisi Coates's influential essay, "The Case for Reparations."[12] Through an intergenerational story of one Black family in Chicago, Coates details the ways racist state housing policies and predatory lending practices deprived Black Americans of one of the, if not the most, significant vehicles of intergenerational wealth: homeownership. This history of "wealth theft," Coates explains, originated with the plunder of life and labor during slavery and continued to the present day with racialized practices such as subprime mortgage lending and the impact of events like the 2008 housing foreclosure crisis.[13] *Reparations Now* picks up on the concept of theft, although in the case of environmental racism, theft is not so much about intergenerational wealth as about the shortening of life: the theft of time, of life, and of relations.

The Movement for Black Lives's *Reparations Now Toolkit*, like the California Reparations Task Force, responds to the failure of the federal government to take up the question of reparations for chattel slavery. This is not for lack of trying. For example, in 1989 Congressional Representative John Conyers introduced H.R. 40, a bill to study reparations for African Americans, to the U.S. Congress. This bill has been reintroduced to Congress every year since then (and, after Conyers's passing in 2019, by Representative Sheila Jackson), but until 2021 it had not made it past the House Judiciary Committee. And so the idea has been increasingly taken up by state and municipal governments.[14] San Francisco's African American Reparations Advisory Committee was established by resolution in February 2020 (prior to the California Task Force) by the city's District 10 supervisor, Shamman Walton, who represents Bayview-Hunters Point. Similar to the scope of the state-level commission, the San Francisco Reparations Advisory Committee includes a health subcommittee to "research, elevate, and provide solutions to address the past, present, and future harm to Black people in San Francisco caused by health disparities, environmental injustices, and the denial of adequate and timely health care."[15] The inclusion of environmental justice concerns within these studies and reports on reparations is arguably due to the work of the many activists and scholars who, especially since the 1980s, have been organizing around and theorizing the connections between racism and environmental harm.

While Ealom had asked for clean air and decent housing as reparations, this does not mean that individual checks or forms of monetary compensation should be off the table. Like the *Reparations Now Toolkit*, the California Task Force relied on the United Nations Principles on Reparations, an international legal framework, to guide its study and recommendations. According to this framework, an effective reparations program must include restitution, compensation, rehabilitation, satisfaction, and guarantees of nonrepetition of harms. The task force worked closely with economists to make precise calculations about the specific, monetary sums owed to descendants of enslaved people and to those descendants who lived in California during specific historical periods (such as bank redlining), which the task force deemed necessary to address the principles of "restitution" and "compensation."[16] However, the task force's report also includes lengthy chapters with specific policy recommendations, thus addressing the principles of rehabilitation and guarantees of nonrepetition of harm. It is the scope and vision of these recommendations I am most interested in, and which speak more closely to Ealom's concept of reparations voiced at the environmental justice demonstration. Namely, were the state of California to adopt all of the report's policy recommendations, we would effectively be living in a different, more just world. Rather, enacting reparations would entail a reimagining of some of society's basic institutions and premises.

Industrial pollution and environmental health are part of these reparations reports, but reparations, in this expansive sense can also contribute to environmental justice scholarship, for example by detailing the historical role of the state in producing racialized environmental harms. In her essay on the water crisis in Flint, Michigan, geographer Malini Ranganthan notes that most critical analyses of the crisis relied on two dominant understandings of environmental racism: racism as an intentional, discriminatory act and race as "correlated" with pollution. Ranganathan argues that both frameworks constrain our thinking on environmental racism, which should, she argues, be understood as inextricable from the workings of the liberal state. She focuses on how the Flint water crisis was the result of state policies leading to housing segregation and allowing for the "conjoined processes of property making and property taking" in twentieth-century U.S. history.[17] This is precisely the connection that recent state and social movement reports on reparations make: that many environmental injustices are part of a longer history of state policies and inaction, stemming from the historical and ongoing dehumanization of Black people in the United States. Moreover, these reparations reports also situate environmental health and justice concerns within a broader vision of social transformation. That is, these reports suggest that environmental justice will not succeed if it is envisioned too narrowly (for example, as simply enhanced regulatory enforcement); it needs to be part of a larger, more radical vision of social change.

Ranganathan turns to the concept of "abolition ecologies" as an analytical framework that emphasizes the ongoing legacies of slavery in producing racialized environments while also emphasizing the creative ways people envision and work toward nonoppressive environments. Because abolition is simultaneously involved in dismantling existing carceral institutions and building more just institutions and ways of relating with each other, it arguably represents a more expansive political horizon that reparations reports—the California and San Francisco reports, after all, seek restitution directly from the state whereas abolition looks beyond the state for justice and healing. Here, philosopher Olúfémi Táíwò's theory of reparations provides an important conceptual bridge between reparations and abolitionist movements. Táíwò advances the notion of a "constructive view" of reparations, which he defines as a "historically informed view of distributive justice, serving a larger and broader world-making project."[18] Táíwò theorizes reparations on the global scale, in response to what he terms "global racial empire." Global racial empire literally constructed the world as we know it, he argues; the constructive view of reparations must be up to the same challenge. It is "neither a project of reconciliation nor redemption" but a "forward-looking target," that of physically—literally—reconstructing the world.[19] Táíwò aligns the constructive view of reparations with the Movement for Black Lives's platform on reparations.[20] The concept of reparations as a redistributive project, moreover, bears similarities with U.S. abolitionist demands, expressed most visibly during the summer of 2020, to defund police departments and reinvest in communities—a demand that was also articulated in terms of redirecting the substantial funding of police departments toward improving environmental protection and amenities for historically divested, polluted neighborhoods.[21]

Reparations can therefore align with, or be part of, abolitionist imaginaries and projects.[22]

Marie Harrison, like many Black Bayview-Hunters Point residents, came to San Francisco from a former slave state in the U.S. South. In the city, she navigated the many afterlives of slavery in racialized housing markets, redevelopment projects, and toxic environments. She organized to change these conditions, in most cases demanding that the state remedy them. At the board of supervisors hearing in 2018, where this book began, Marie told the elected officials: "I get angry when I see a three- or four-year old with asthma. Is that by design for our community?" Five years later, after Marie, a nonsmoker, had succumbed to chronic lung disease, the California Reparations Task Force could be seen to answer her question. As the report states: "The harms African Americans have experienced have not been incidental or accidental—they have been by design."[23] Constructing the world Marie and other residents have envisioned, over many decades, will most likely not come solely from petitioning the state, or, for that matter, from "inside liberalism," as Ranganathan puts it.[24] That does not mean that people should stop trying to strengthen and improve state practices, only that these efforts should be part of a broader, liberatory politics. In this book I have tried to show how Bayview-Hunters Point residents articulated visions of social change, often in state spaces—the point, for me, is the presence and persistence of an oppositional interpretive framework and at times a radical imaginary, across generations. The political campaigns and visions examined in this book thus serve as a reminder that "in the poetics of struggle and lived experience, in the utterance of ordinary folk, in the cultural products of social movements, in the reflections of activists, we discover the many different cognitive maps of the future, of the world not yet born."[25]

Notes

INTRODUCTION

1. Bishari, "Two Imprisoned for Hunters Point Shipyard Cleanup Fiasco."

2. *Remediation* is the technical term for what is colloquially called "cleanup" at hazardous waste sites.

3. In 2010 the Hunters Point Shipyard Redevelopment Plan was amended to include Candlestick Point, which also includes the former Candlestick Park Stadium.

4. In their essay, "Fugitive Justice," Saidiya Hartman and Stephen Best write about the injuries and afterlives of slavery, the impossibility of redress, and fugitive forms of justice. They discuss an eighteenth-century, formerly enslaved writer, Ottobah Cugoano, who writes of the necessity for legal remedy for the injustices of slavery, even as he also acknowledges any such remedy would be insufficient. "Cugoano shuttles between grief and grievance. That is, he demands justice in light of that which he cannot describe or convey, fully cognizant that what has been destroyed cannot be restored." Best and Hartman, "Fugitive Justice," 2. Marie could also be thought to petition the state to remedy what is in many ways socially and ecologically irreparable.

5. On market-oriented sustainability, see Greenberg, "What on Earth Is Sustainable?"

6. On postindustrial greening, see Lang and Rothenberg, "Neoliberal Urbanism, Public Space"; Safransky, "Greening the Urban Frontier"; and Wachsmuth and Angelo, "Green and Gray."

7. Greenberg, "Sustainability Edge."

8. By "urban restructuring" I refer to long-term, widespread changes in the built environment, underlying political economy, and demographics of this area of the city.

9. Fuller, "Loneliness of Being Black in San Francisco"; and *Report of the San Francisco Mayor's Task Force on African-American Outmigration*. The idea of Bayview-Hunters Point as the last Black neighborhood in San Francisco, albeit threatened by gentrification and Black displacement, is part of the narrative of the poignant 2019 film, *The Last Black Man in San Francisco*.

10. Krupar, *Hotspotters Report*, 3.

11. On "slow violence," see Nixon, *Slow Violence*.

12. McElroy, "Postsocialism and the Tech Boom 2.0"; and Walker, "Landscape and City Life."

13. Maharawal, "Protest of Gentrification." See also McElroy, "Postsocialism and the Tech Boom 2.0"; and Walker, *Pictures of a Gone City*.

14. Darling et al, "Just Biomedicine on Third Street?"

15. San Francisco Office of the Assessor, *Annual Reports*, 2006–2022.

16. McCormick, "Bayview's Black Exodus"; Beckett, "Aboriginal Blackmen United"; and Hua, "Not Business as Usual."

17. Markusen, *Rise of the Gunbelt*; and Sorenson, *Shutting Down the Cold War*.

18. Sorenson, *Shutting Down the Cold War*.

19. SF Parks Alliance, Blue Greenway. On turning brownfields into green spaces, see Checker, "Wiped Out by the 'Greenwave'"; and Anguelovski, "Parks as (Green) LULUs?"

20. Angleo, *How Green Became Good*.

21. On "green" or "environmental" gentrification, see Checker, *Sustainability Myth*; Gould and Lewis, *Green Gentrification*; Safransky, "Greening the Urban Frontier"; Anguelovski et al., "Expanding the Boundaries of Justice in Urban Greening Scholarship"; and Curran and Hamilton, "Just Green Enough."

22. As one indication, since 2006, Bayview-Hunters Point consistently led the city in assessed property value growth, often a greater than 12.5 percent growth rate, largely triggered by the new light rail line and the promise of redevelopment at the shipyard. San Francisco Office of the Assessor, *Annual Reports*, 2006–2022.

23. On urban renewal in the Fillmore, see Pepin and Watts, *Harlem of the West*; Mollenkopf, *Contested City*; Lai, "Racial Triangulation of Space"; Allensworth, Hall, and McElroy, "(Dis)location/Black Exodus"; Brahinsky, "'Hush Puppies'"; and Jackson and Jones, "Remember the Fillmore."

24. Cebul, "Tearing Down Black America."

25. Patricia is a pseudonym. Pseudonyms are used throughout the book unless individuals consented to be quoted by their real name. Some of my interlocuters had passed away at the time of publication, and I have used pseudonyms in their cases as well.

26. McCormick, "Bayview's Black Exodus."

27. Selna, "Campaign 2006." According to District 10 (which includes Bayview-Hunters Point) supervisor Sophie Maxwell, the petition misrepresented the redevelopment plan to signatories of the referendum campaign, including false claims about eminent domain. Even though the referendum was not admitted onto the city ballot, four city supervisors (not including Maxwell) voted against the redevelopment plan, citing concerns about democratic process. See Jones, "Battle for Bayview."

28. Louis is a pseudonym.

29. Baranski, *Housing City by the Bay*.

30. Horowitz, *Katrina*; Gotham and Greenberg, *Crisis Cities*; and Tompkins, "There's No Bad Teacher."

31. On accumulation by dispossession, see Harvey, *New Imperialism*.

32. McGurty, *Transforming Environmentalism*.

33. U.S. Government Accountability Office, *Siting of Hazardous Waste Landfills*.

34. *Principles of Environmental Justice*.

35. Pulido and De Lara, "Reimagining 'Justice,'" 77.

36. Pellow, "Black Lives Matter"; Pellow, *Critical Environmental Justice*; Ranganathan, "Thinking with Flint"; Pulido, "Geographies of Race and Ethnicity II"; and Sze, *Environmental Justice*. See also Kurtz, "Acknowledging the Racial State."

37. Pulido, "Geographies of Race and Ethnicity II"; see also Fabricant, *Fighting to Breathe*.

38. Pulido, Kohl, and Cotton, "State Regulation and Environmental Justice," 15; see also Konisky, *Failed Promises*.

39. Harrison, *From the Inside Out*.

40. Pellow, "Black Lives Matter."

41. Pulido, Kohl, and Cotton, "State Regulation and Environmental Justice," 12.

42. Pulido, Kohl, and Cotton, "State Regulation and Environmental Justice," 27.

43. Thompson, "Toward a World Where We Can Breathe," 9.

44. Dhillon, "Indigenous Resistance, Planetary Dystopia"; Gilio-Whitaker, *As Long as Grass Grows*.

45. On organizations "mingl[ing] reformist and radical ideologies and strategies," see Gilmore, *Golden Gulag*, 186.

46. Perkins, *Evolution of a Movement*, 6; and Harrison, "Environmental Justice and the State."

47. Woods, *Development Arrested*, 39.

48. Murphy, "Alterlife," 495–496. See also Hepler-Smith, "Molecular Bureaucracy"; Liboiron, Tironi, and Calvillo, "Toxic Politics"; and Ahmann and Kenner, "Breathing Late Industrialism."

49. Liboiron, Tironi, and Calvillo, "Toxic Politics," 333.

50. Murphy, "Alterlife," 496.

51. "New Gardens," 8.

52. Lelchuk, "City Confirms Workers' Charges."

53. See "Hunters Point Shipyard Residents"; Hepler, "New Legal Challenge"; Fagone and Dizikes, "Working in a Wasteland"; and Wolfrom, "Fresh Concerns Raised."

54. In 1980, Bayview-Hunters Point was 72 percent Black. Today, the demographic profile is more varied: 27 percent Black, 24 percent Latinx, and 37 percent Asian and Pacific Islander. U.S. Census data cited in San Francisco Municipal Transportation Agency, *Bayview Transportation Plan*. See also *Report of the San Francisco Mayor's Task Force on African-American Outmigration*.

55. Other scholars have written incisively about southeast San Francisco, however. See Brahinsky, "Race and the Making of Southeast San Francisco"; Dunn-Salahuddin, "Forgotten Community"; Shange, *Progressive Dystopia*; and Kang, "Respect for Community Narratives."

56. Bernice is a pseudonym.

57. TallBear, "Standing with Faith."

58. Táíwò, *Reconsidering Reparations*.

1. THE WASTELANDING OF SOUTHEAST SAN FRANCISCO

1. Anonymous, "Chinese Shrimpers," quoted in Brahinsky, "Making and Unmaking of Southeast San Francisco," 52. On Chinese fishing encampments in southeast San Francisco, see McEvoy, *Fisherman's Problem*; and Hamusek-McGann, Blossom, and Maniery, *Archeological Inventory and Assessment*.

2. Brechin, *Imperial San Francisco*; Walker, "Industry Builds the City"; Igler, *Industrial Cowboys*; and Robichaud, *Animal City*. The industrial history of San Francisco in this chapter is also derived from my own study of Sanborn Insurance Maps and San Francisco Business Directories, located at the Bancroft Library at UC Berkeley, from the nineteenth and early twentieth centuries.

3. Saxton, *Indispensable Enemy*. There is another connection between Chinese San Franciscans and Hunters Point. In the late nineteenth and early twentieth centuries, San Francisco businessmen periodically sought to relocate Chinatown (which was and still is adjacent to the city's central business district) to Hunters Point. These plans for relocating Chinatown to Hunters Point became so popular that in May 1906 a Chinese diplomat from Washington, D.C., traveled to San Francisco to intervene. "Suggests a New Chinatown."

4. Voyles, *Wastelanding*.

5. Voyles's book centers on an historical analysis of the discursive and material wastelanding of Diné territory, and she argues that settler colonialism depended on a racialized production of Indigenous lands as empty and wasted, requiring Anglo settlement. Although Indigenous identity is not a racial or ethnic category, it is important to recognize, as Cheryl Harris notes, "Indigeneity was and is racialized." Harris, "Of Blackness and Indigeneity."

6. Pulido, "Geographies of Race and Ethnicity II"; and Gilmore, "Abolition Geography and the Problem of Innocence."

7. Laura Pulido's work is instrumental here, especially Pulido, Sidawi, and Vos, "Archaeology of Environmental Racism"; and Pulido, "Rethinking Environmental Racism."

8. Quoted in Blomley, *Unsettling the City*, 110. As historian Kyle May argues, modern U.S. cities were constructed through the dispossession of both Indigenous and Black people. Mays uses an analytic of dispossession and forms of resistance and counternarratives to it as one way to understand "the connection between racial capitalism and settler colonialism, as well as indigeneity and racialization in urban spaces." Mays, *City of Dispossessions*.

9. Hawthorne and Lewis, "Black Geographies," 10.

10. Moore, *Take This Hammer*.

11. Booker, *Down by the Bay*, 32.

12. Anderson, *Tending the Wild*, 1.

13. Wynter, "1492."

14. Lightfoot et al., "European Colonialism and the Anthropocene," 106. See also Anderson, *Tending the Wild*; and Akins and Bauer, *We Are the Land*.

15. According to Field et al, "'Ohlone' is a contemporary term, grouping together all the descendants of all the indigenous peoples who lived in the entire Bay Area, extending inland from the Carquinez Strait south to Soledad, and from the Golden Gate to Big Sur on the Pacific Coast." Field et al., "Contemporary Ohlone Tribal Revitalization Movement." On Indigenous land use in the San Francisco Bay Area and in southeast San Francisco, see Booker, *Down by the Bay*; Luby, Drescher, and Lightfoot, "Shell Mounds and Mounded Landscapes"; and Olmsted, *Rincón de las Salinas*.

16. Matsuda, *Pacific Worlds*; and Wheeler, "Empires in Conflict and Cooperation."

17. Brown, *Mining in Latin America*. Or, as Edward Galeano writes, "The metals taken from the new colonial dominions not only stimulated Europe's economic development; one may say that they made it possible." Galeano, *Open Veins of Latin America*, 23.

18. Lightfoot et al., "European Colonialism and the Anthropocene"; and Engstrand, "Seekers of the 'Northern Mystery'".

19. Crosby, *Ecological Imperialism*.

20. Lightfoot et al., "European Colonialism and the Anthropocene," 107. See also Hackel, "Land, Labor, and Production"; and Crosby, *Ecological Imperialism*. Indeed, these environmental changes, especially from livestock overgrazing lands, pushed more Indigenous people onto the missions. See Akins and Bauer, *We Are the Land*.

21. Hurtado, *Indian Survival*; and Garone, *Fall and Rise of Wetlands*.

22. Akins and Bauer, *We Are the Land*, 80.

23. Also known as Mission Dolores, today this is the location of San Francisco's Mission District.

24. Olmsted, *Rincón de las Salinas*; and Robichaud, *Blood in the Water*.

25. Hackel, "Land, Labor, and Production"; and Hornbeck, "Land Tenure and Rancho Expansion."

26. Hackel, "Land, Labor, and Production," 132.

27. Madley, *American Genocide*, 52–53.

28. Reich, "Dismantling the Pueblo"; Olmsted, *Rincón de las Salinas*; and Robichaud, *Animal City*.

29. Horsman, *Race and Manifest Destiny*; Brechin, *Imperial San Francisco*; and Igler, *Industrial Cowboys*. As a side note, Karl Marx and Fredrich Engels supported the U.S. in its 1846–1848 war with Mexico. Horace Davis notes that these two critics of capitalism saw the Mexican rancho economy in Alta California as backward and feudal and believed the U.S. annexation of California would more quickly lead to the proletarianization of the region. Cedric Robinson draws on Davis's writings on the racialized perception of Mexicans held by Marx and Engels, and their support for an imperial war, in developing his critique of Western Marxism for its theoretical inability to recognize the racial character of capitalism. See Davis, "Nations, Colonies and Social Classes"; and Robinson, *Black Marxism*.

30. Montoya, *Translating Property*; and Correia, "Making Destiny Manifest."

31. During the hearing, U.S. officials interrogated San Francisco residents to ascertain whether Bernal had in fact improved his property (a requirement of ownership under Mexican property law). Lawyers for the board of commissioners asked questions such as these: "To what extent did the cattle range on the Rancho? How many of them were there, and how large was the garden?" "Transcript of Proceedings in Case No. 30."

32. It is unclear, from the available historical record, why Bernal sold part of his land grant to real estate speculators, but it is likely he hoped to benefit from the increased economic value of his land after 1848. As the historian David Igler notes, "Contrary to the romantic myth of pastoral Mexican California, the Californio elite had also actively participated in market-based trade and developed their lands as private property." *Industrial Cowboys*, 45.

33. On the new title holders to Hunters Point see *Bayview-Hunters Point Area B Survey*.

34. Hamusek-McGann, Baker, and Maniery, *Archeological Inventory and Assessment of Hunters Point Shipyard*; *Bayview-Hunters Point Area B Survey*;

Olmsted, *Rincon de las Salinas*; Robichaud, *Animal City*; and O'Brien, *San Francisco's Bayview Hunters Point*.

35. Dow, "Bay Fill in San Francisco"; Booker, *Down by the Bay*; and Dillon, "Civilizing Swamps in California."

36. Walker, "Industry Builds the City."

37. Brechin, *Imperial San Francisco*; and *Crocker and Langley Business Directory*, 2,163.

38. Takaki, *Iron Cages*; and Walker, "Industry Builds the City."

39. Igler, *Industrial Cowboys*.

40. Brechin *Imperial San Francisco*; Dow, "Bay Fill in San Francisco"; and Henderson, *California and Fictions of Capital*. The *Daily Alta California* commented on the importance of the shipyard for the city's ambitions: "Thus step by step our capitalists are providing all that it wanting [*sic*] to secure to our people a full realization of their most sanguine hopes in behalf of the Empire City of this Coast." "Stone Dry Dock."

41. Dow, "Bay Fill in San Francisco."

42. "Southern Suburbs."

43. "Bayview to Have More Homes."

44. *Bayview-Hunters Point Area B Survey*.

45. Maantay, "Zoning, Equity, and Public health"; and Weiss, "Real Estate Industry."

46. Safransky, *City after Property*.

47. San Francisco Homeowners Loan Corporation residential security map. Nelson et al., "Mapping Inequality."

48. Louis imagined his journey to San Francisco, as a young boy, around the time the artist Jacob Lawrence painted *The Migration of the Negro* series (1939–1940) at his studio in Harlem. Lawrence's sixty captioned panels depict the Great Migration—the mass relocation of African Americans from the rural South to the urban and industrial North. Lawrence read novels, researched clipping files, and listened to oral histories of southern migrants at the 135th Street branch of the New York Public Library in preparation for his paintings. Collectively, *The Migration* paintings tell a story of how "the Negro, who had been part of the soil for many years, was now going into and living a new life in the urban centers" (caption for Panel 7). Some of Lawrence's paintings depict a sense of hope—bodies tilt toward one side of the painting frame, with a clear sense of direction. Lawrence's paintings also depict new forms of racism Black migrants navigated in the urban North. See Capozzola, "Jacob Lawrence."

49. Du Bois, "Colored California."

50. Broussard, *Black San Francisco*; and Daniels, *Pioneer Urbanites*.

51. Even before the United States officially entered the war, the federal government spent $60 million on defense facilities in the Bay Area to manufacture ships for the Lend-Lease Program in England. Scott, *San Francisco Bay Area*.

52. Moore, *To Place Our Deeds*; Johnson, *Second Gold Rush*; and Jelly-Schapiro, "High Tide, Low Ebb."

53. Broussard, *Black San Francisco*; and Johnson, *Negro War Worker*.

54. Allen, *Port Chicago Mutiny*; and Arbona, "After the Blast."

55. Fifty of those men were subsequently convicted of mutiny by an all-white tribunal. They were exonerated in 2019. See Allen, *Port Chicago Mutiny*.

56. Arbona, "Anti-Memorials and World War II," 183.

57. Washburn, "Pittsburgh Courier's Campaign"; and Washburn, *African American Newspaper*.

58. Quoted in Washburn, *African American Newspaper*.

59. Broussard, *Black San Francisco*.

60. Johnson received his doctorate in sociology at the University of Chicago, a renowned center of urban sociology. *The Negro War Worker in San Francisco* was funded by the local Y.W.C.A., the Race Relations Division of the American Missionary Association, and the Rosenwald Fund. Miller, *Postwar Struggle for Civil Rights*.

61. Johnson, *Negro War Worker*, 29.

62. Although overcrowding was certainly a function of the restrictions Black migrants faced in securing housing, Johnson also considered it a form of mutual aid: a form of solidarity and communal support.

63. Quoted in Broussard, *Black San Francisco*, 174.

64. Baranski, *Housing the City by the Bay*; and Johnson, *Negro War Worker*. San Francisco built approximately 10,000 temporary dwelling units for military workers throughout the city.

65. Broussard, *Black San Francisco*.

66. Johnson, *Negro War Worker*, 31.

67. Johnston, *Half-Lives and Half-Truths*; Weisgall, "Nuclear Nomads of Bikini"; and Gusterson, *People of the Bomb*. When the U.S. military descended on Bikini in 1946, it convinced native Bikinians to relocate to the island of Rongerik, 140 miles away, with the promise of returning after Operation Crossroads was over. But the U.S. military's bombs made their return impossible; meanwhile, Bikinians starved on Rongerik. In 1948 the United States moved Bikinians again, first to Kwajalein and then to Kili island, 400 miles south of Bikini. Once a self-sufficient people, on Kili, Bikinians grew dependent on U.S. food aid. Bikini remains uninhabited.

68. *Radiation Exposure from Pacific Nuclear Tests*.

69. "Operations Crossroads, 1946"; and Weisgall, *Operation Crossroads*.

70. Miller, *Under the Cloud*; and Weisgall, *Operation Crossroads*.

71. Hirsch, "EPA to Hunters Point"; and Bradley, *No Place to Hide*.

72. U.S. Department of the Navy, Base Realignment and Closure, "Final Historical Radiological Assessment."

73. "Topics for Investigation by Radiation Laboratory." NRDL biomedical research later helped establish national standards for "permissible" radiation

exposure. In fact, when President Eisenhower established the Federal Radiation Council (FRC) in 1959, he appointed the NRDL's scientific director, Paul Tompkins, as executive director of the new agency. The NRDL also provided key testimony at a congressional hearing in 1960 that laid the basis for federal radiation standards. Ten years later, in 1970, the FRC was folded into the newly established EPA. Today, in a strange twist, NRDL expertise comes back to the Hunters Point Shipyard in the guise of (contested) radiation safety standards for Superfund remediation. Jones, "Review of U.S. Radiation Protection Regulations"; and Cram, "Becoming Jane."

74. "Appendix B: Interviews," in U.S. Department of the Navy, Base Realignment and Closure, "Final Historical Radiological Assessment." The lab built such elaborate animal testing facilities on the Hunters Point Shipyard that when the shipyard closed in 1974, the National Institutes of Health and University of California, San Francisco, vied to acquire them. The NRDL also operated a joint animal colony in the Marshall Islands with the Naval Medical Research Institute, breeding animals that were used during nuclear weapons tests. Norbert P. Page to John A Sutro.

On the NRDL's involvement with studying the biomedical impacts of the 1954 "Castle" bomb on Marshall Islanders (also known as Project 4.1), see Harkewicz, "Ghost of the Bomb."

75. Annual Report, U.S. Naval Radiological Defense Laboratory, 1953.

76. "US NRDL History 1946–1955."

77. "Monthly Progress Report, Radiation Laboratory, for Period Ending 31 March 1947."

78. In August 2001, Janice Gale, a former librarian at the NRDL, visited the Hunters Point Naval Shipyard Restoration Advisory Board's (RAB) monthly meeting. Gale told the RAB about three malignant tumors on her carotid artery, which she connected to the lack of health and safety regulations for NRDL workers in the early years of the lab. Gale recalled starting her job at the NRDL in April 1948, and "in September all of a sudden we had a badge and a Geiger counter so that when we left in the afternoon, we would pass the Geiger counter. But prior to that time, there was nothing." While significant in themselves—to Gale and her family—these details also suggest that the navy's and NRDL's radiological practices were likely ad hoc and haphazard, at least initially, and that the official historical record of radioactive waste practices at the shipyard represents an incomplete accounting of the past. Hunters Point Restoration Advisory Board, Reporter's Transcript of Meeting, August 23, 2001.

79. "Request for Approval to Construct Isotope Storage Building."

80. According to its own records and interviews later conducted with lab workers, the NRDL released liquid radioactive waste into the shipyard's sewer system, disposed of irradiated animals on the base, and packed other radioactive waste products into fifty-five-gallon steel drums, reinforced with concrete, so they would sink to the bottom of the ocean. In fact, the Hunters Point Naval Shipyard

became a center of waste disposal for radioactive waste generated by laboratories and military installations around the Bay Area and oversaw this work until waste disposal was contracted out to a private company in 1959. In 1953, for example, ten barge trips with an average of 5,800 gallons of radioactive waste from NRDL and other installations left Hunters Point; between 1954 and 1958, the NRDL sank 3,762 waste drums at sea. In 1980, the EPA estimated that 25 percent of the NRDL's waste drums likely broke under water pressure before they settled on the ocean floor. U.S. Department of the Navy, Base Realignment and Closure, "Final Historical Radiological Assessment," Appendix B: Interviews; Davis, "Fallout"; and "Washington Notes."

81. U.S. Environmental Protection Agency, "Hunters Point Naval Shipyard Superfund Site Profile."

82. On "industrial overburden" see Fabricant, *Fighting to Breathe.*

83. San Francisco Department of City Planning, *South Bayshore, 1970 Census.*

84. Self, *American Babylon*; Massey and Denton, *American Apartheid*; and Lipsitz, *Possessive Investment in Whiteness.*

85. Baranski, *Housing in the City.* The SFHA's first three permanent projects, constructed in the 1940s (Holly Court, Sunnydale, and Potrero), were only open to white tenants.

86. Quoted in Broussard, *Black San Francisco*, 172. Beard's tenant selection policy was not idiosyncratic; rather, it reflected New Deal–era national housing policy. A decade earlier, Public Works Administration director Harold Ickes had instituted a similar "neighborhood composition rule," preventing government housing projects from altering the racial composition of neighborhoods. Hirsch, *Making the Second Ghetto.*

87. Baranski, *Housing in the City.*

88. Broussard, *Black San Franscisco.*

89. In his study of housing discrimination in Chicago, Arnold Hirsch refers to this as "making the second ghetto." According to Hirsch, the "most distinguishing feature" of the post–World War II "ghetto expansion" was "that it was carried out with government sanction and support." See Hirsch, *Making the Second Ghetto*, 9.

90. Fainstein, *Restructuring the City.* The postwar years also saw a steep decline in the city's maritime industry. When the United States entered World War II in 1942, San Francisco's port system accounted for the largest share of general cargo passing through the Bay Area, at 63.2 percent. By 1957, this share had dropped to 26.7 percent, largely as a consequence of new cargo-handling and shipping technologies that favored the more expansive backlots of the Port of Oakland. In response, the Port of San Francisco began to consider other uses for the city's waterfront, such as leasing port land to investors and encouraging the development of office buildings, apartments, and marinas. That is, the transformation of San Francisco's waterfront into a space of commerce and leisure

began in the late 1950s. See Rubin, *Negotiated Landscape*; and Ebasco Services, *Port of San Francisco*.

91. Razing and redeveloping the Fillmore/Western Addition was a stated goal of the Bay Area Council (BAC). This consortium of local business leaders (including representatives from Bank of America, Standard Oil, Pacific Gas and Electric, and Bechtel Corporation) aimed to promote regional economic development, working in part through the built environment. For example, BAC was the primary planner and lobbyist for the Bay Area Regional Transit (BART) system. In 1956, several BAC members formed the Blyth-Zellerbach Committee, which lobbied for urban renewal projects to reinvigorate San Francisco's central business district, including the Fillmore/Western Addition. Hartman and Carnochan, *City for Sale*.

92. Mollenkopf, *Contested City*.

93. Pepin and Watts, *Harlem of the West*; and Allenworth, Hall, and McElroy, "(Dis)location/Black Exodus."

94. Brahinsky, "'Hush Puppies.'"

95. Fainstein, *Restructuring the City*.

96. *Commercial and Industrial Activity*.

97. One some of the affective dimensions of racialized industrial toxicity, see Vasudevan, "Intimate Inventory of Race and Waste."

98. Lewis, *Violent Utopia*, 49.

99. On wastelands as new accumulation frontiers, see Knuth, Potts, and Goldstein, "In Value's Shadows"; and Dillon, "Race, Waste, and Space."

100. Selna, "News/Politics."

101. In his theory of gentrification, Neil Smith identifies two related economic processes driving the return of capital to "inner cities" and the subsequent displacement of working-class residents, which began in earnest in the 1970s in global financial centers like New York City and London. First, the emergence of a *rent gap*, or "the disparity between the potential ground rent level and the actual ground rent capitalized under the present land use," attracts investors and individual homebuyers, because they can purchase real estate cheaply, pay rehabilitation costs, and eventually sell the property, making a profit. Second, and relatedly, Smith points to the contradictory tendencies of uneven development under capitalism, and in particular the tendency toward spatial differentiation. In the post–World War II decades, the most marked spatial differentiation on the urban scale was the flight of capital to the suburbs and the related, systemic devalorization of the inner city. And yet, Smith writes, this devalorization later became the basis for new investment opportunities, in the form of gentrification. Sarah Knuth, Shaina Potts, and Jennifer Goldstein build on this aspect of Smith's thesis, pointing to the many forms of devalorization, such as land degradation, which have in recent years become an "accumulation frontier." Similarly, I see wastelanding in southeast

San Francisco as producing a *toxic rent gap*, or a value differential produced by industrial and military toxicity, which becomes the basis for new rounds of capital accumulation. Smith, *New Urban Frontier*; and Knuth, Potts, and Goldstein, "In Value's Shadows."

2. BLACK COUNTERPLANNING
FOR A NEW HUNTERS POINT

1. The Area Development office on Navy Road, one of four antipoverty "outposts" of the Hunters Point Area Development Program, was authorized and funded by the federal Economic Opportunity Act of 1964. The legislation was part of Johnson's broader War on Poverty agenda. The decentralized structure of antipoverty offices in Hunters Point reflected the community action goal of "maximum feasible participation of residents of the areas and members of the groups served." See Kramer, *Participation of the Poor*, 1; and "New Anti-Poverty Outpost."

2. "Outbreak"; and *128 Hours*. For critical accounts of the uprising see Crowe, *Prophets of Rage*; Agee, *Streets of San Francisco*; and Dunn-Salahuddin, "Forgotten Community." Alvin Johnson, the police officer who killed Matthew Johnson, later received the verdict of "excusable homicide" in court. As a historical side note, Matthew Johnson's mother was represented by John Lee Cochrane Jr., a Los Angeles–based attorney who had represented the widow of Leonard Deadwyler, a Black man killed by the Los Angeles Police Department in May 1966. Later, in the 1990s, Cochrane represented O. J. Simpson. "Cop Goes Free."

3. Agee, *Streets of San Francisco*.

4. Dunn-Salahuddin, "Forgotten Community," 225.

5. *The Spokesman* was established by Black Hunters Point residents and was published from 1965 to 1969, supported by federal funding. Another newspaper titled *The Spokesman* had been published in San Francisco between 1930 and 1936. That paper was founded and published by John Pittman, who received his undergraduate degree from Morehouse and an MA in economics from UC Berkeley. Pittman was involved in the Communist Party in San Francisco and later published *People's World*. See Broussard, *Black San Francisco*.

6. Dunn-Salahuddin, "Forgotten Community."

7. "Outbreak." Hunters Point resident and community leader Osceola Washington also made this point during a Senate hearing on the War on Poverty, held in San Francisco, on May 10, 1967. Senator Edward Kennedy had asked Washington, referring to the uprising: "You had problems about a year ago—violence, disorder, lawlessness. Have the conditions that brought about these problems changed over the period of the last twelve months?" Washington responded:

"This did not happen overnight; this is something that has been handed down from 1954 when I came to San Francisco. The young people that were in this incident were born about that time. These young men have grown up in this situation: police brutality, poor schools, poor educations, kickouts, mostly—they called them dropouts but they are kickouts from Hunters Point. . . . So this is something that has grown up in the young people; this hostility is something that did not happen overnight." Washington stresses the structural exclusions and forms of state violence experienced by young people in Hunters Point. She attempts to correct Senator Kennedy's assumption that the underlying conditions of the 1966 uprising could be easily corrected over the course of one year, implying that much more deep-seated, structural changes were needed. *Examination of the War on Poverty.*

8. "'Plant in' at Death Site."

9. McKittrick. *Demonic Grounds.* One of the outcomes of the 1966 uprising was the influential Hunters Point organization, Young Men for Action (YMA). YMA leadership came from an older organization working in Hunters Point, Youth for Service (YFS), founded in 1957 by a white Quaker, Carl May, with funding from the American Friends Service Committee. YFS organized volunteer activities for Hunters Point teenagers (exclusively boys), such as painting houses and tending gardens, and was later funded by the Ford Foundation. After the uprising, YFS youth and Luster were celebrated in the local press for their "peacekeeping" efforts during the uprising, and some of these young men formed a new group, YMA. Adam Rogers (born in San Francisco in 1944; he grew up on the hill) participated in YFS programs and became a prominent figure during the uprising through his efforts to mediate between the police and those demonstrating in the streets; he later became a leader of YMA. Rogers and YMA subsequently worked closely with Mayor Joseph Alioto (who succeeded John Shelley), who provided YMA workers jobs in exchange for their political support. Rather than attempting to describe the entirety of political action in Hunters Point, this chapter focuses primarily on Black women at the center of what I consider "counterplanning" around housing, health, and food. On Luster, Rogers, and other prominent Black men in Bayview-Hunters Point's historical political landscape, see Agee, *Streets of San Francisco.*

10. McKittrick, *Demonic Grounds.*

11. Gilmore, *Golden Gulag*, 179.

12. Importantly, not all Bayview-Hunters Point residents had the same vision. The meager funds available through the San Francisco Economic Opportunity Council and the Model Cities Act, along with Mayor Alioto's divide-and-conquer style of politics led to internal conflicts over priorities, principles, and leadership.

13. "Untitled Document."

14. McKittrick, *Demonic Grounds*, 17.

15. "Residents Stop Eviction."

16. "Residents Stop Eviction."

17. "Should Do Something for Neighborhood."

18. Baranski, *Housing the City by the Bay.*

19. "Alice Griffith Projects."

20. Taylor, *Race for Profit*, 26–28.

21. Taylor, *Race for Profit*; and McLaughlin, "Pied Piper of the Ghetto."

22. "Bayview Community Speak Out."

23. Kramer, *Participation of the Poor*, 49; and "Declaring Policy Affecting Housing Projects." The wartime temporary housing units were transferred to the SFHA after the war. By 1961, the SFHA had concentrated 42 percent of the city's public housing stock in Hunters Point. See Agee, *Streets of San Francisco.*

24. Thompson, "Redevelopment Game!"

25. "Rent Strike."

26. "Housing Commission Forced to Listen."

27. Baranski, *Housing the City by the Bay.*

28. "Hundreds Join Rent Strike." The HPTU also sought legal counsel from the San Francisco Neighborhood Assistance, a federally funded organization.

29. Baranski, *Housing the City by the Bay*, 130.

30. "George Earl on Tenants Union."

31. Baranski, *Housing the City by the Bay*, 130.

32. Wanzer-Serrano, *New York Young Lords*; and Gandy, *Concrete and Clay.*

33. Fredericks, *Garbage Citizenship.*

34. On reproductive labor under conditions of slow violence, see Cairns, "Caring about Water in Camden."

35. Gilmore, *Golden Gulag*, 183.

36. These groups included the United San Francisco Freedom Movement (UFM), a confederation that included the NAACP, CORE, and the Ad Hoc Committee to End Discrimination. Many of these organizations and activists were also engaged in challenging the San Francisco Redevelopment Agency's urban renewal project in the Fillmore. Ralph Kramer writes that the leaders of the UFM later played an important role in the struggle with Mayor John Shelley for control of the Economic Opportunity Council, the city's main War on Poverty office. Kramer, *Participation of the Poor*; Wirt, *Power in the City*; Becker and Myhill, *Power and Participation*; and Miller, *Postwar Struggle for Civil Rights.*

37. Self, *American Babylon*, 183.

38. Brahinsky, "'Hush Puppies.'"

39. Kramer, *Participation of the Poor*, 1.

40. O'Connor, *Poverty Knowledge*; Roy, Schrader, and Crane, "'Anti-Poverty Hoax'"; and Immerwahr, *Thinking Small.*

41. Mollenkopf, *Contested City*; and Gregory, *Black Corona.*

42. Self, *American Babylon*, 203.

43. "Large Crowd Mourns Youth"; and Seale, "Black Panthers and Hunters Point."

44. "How Block Clubs Work."

45. "Portrait of George Earl." In its January 20, 1966, issue, the *Spokesman* listed the Inter-Block Council, the Area Planning Board, and the Bayview-Hunters Point Community Development Corporation as examples of how "committed people . . . have been effective in permeating and shaking the very foundation of the established order." "Bayview an Organized Community."

46. Self, *American Babylon*, 233. When the Community Action Program is discussed by critical social scientists today, it is often dismissed as a state tactic to contain and incorporate radical protest. However, I worry that this critique, in some cases, skips too quickly over the ways local activists appropriated the language of community action and used federally funded antipoverty institutions for their own, locally determined projects and goals.

47. "M. Justin Herman to Harry D. Ross."

48. Fainstein, *Restructuring the City*, 223.

49. "Declaring Policy Affecting Hunters Point."

50. "Alioto Pledge on Hunters Point."

51. "Group Demands Better Housing."

52. "Hunters Pointers Want to Stay."

53. "Portrait of Osceola Washington."

54. "Residents Start Housing Co-op."

55. "Men Take Over." The CDC was federally funded, through the Hunters Point Area Planning Board, though the board retained control over how federal funds were spent in the neighborhood. See Kramer, *Participation of the Poor*.

56. "Group Demands Better Housing."

57. "Planning Brd Firm with Herman."

58. Ralph Kramer writes that through the JHC, Hunters Point residents "were able to extract an unusual concession in getting the redevelopment agency to agree to erect new low-cost housing under the nonprofit corporation's sponsorship before any demolition would take place." *Participation of the Poor*, 53.

59. "BVHP Committee Constitution."

60. "Property vs. Human Rights."

61. Alexander von Hoffman explains how federal housing policy in the late 1960s shifted toward emphasizing the role of the private sector in building and maintaining social housing projects. However, Robert Weaver (the top U.S. housing official under Presidents Kennedy and Johnson) had intitially emphasized the role of nonprofit and cooperative housing developers, prior to adopting an "enthusiasm for funding businesses . . . to implement social welfare policy." The JHC-SFRA's vision for a new Hunters Point reflected this earlier emphasis on nonprofit-sponsored housing. "Genius of Private Enterprise," 166.

62. "Description of Hunters Point."

63. Canter, "Hunters Point Ownership in Question."

64. "Property vs. Human Rights."

65. "Hunters Point Dissidents Want Action."

66. "Lillian Woods to Joseph L. Alioto, M. Justin Herman, and Robert B. Pitts." As an indication that Woods was quite aware of the limitations of the SFRA and the various levels of authority on the issue of redevelopment in Hunters Point, on September 6, 1968, she wrote to President Johnson: "We, the stepchildren of San Francisco the second-class citizens of the 'great' state of California and the United States of America, will not stand for being pawns in this new political chess game . . . now being played out between the City of San Francisco, the present administration of the State of California, and the Department of Housing and Urban Development." Woods's letter demanded that the three government bodies honor their previously stated commitments to redevelopment in Hunters Point.

67. "Lillian Woods to Joseph L. Alioto, M. Justin Herman, and Robert B. Pitts."

68. Hunters Point & India Basin Industrial Park Calendar.

69. Papazoglakis, "'Feminist, Gun-Toting Abolitionist.'"

70. "History of the Neighborhood Co-op."

71. Reese, *Black Food Geographies*, 8.

72. "Character of Mrs. Essie Webb."

73. "Untitled Document."

74. Coleman, "Hunters Point-Bayview Health Service."

75. "Untitled Document"

76. "Untitled Document"; and "To Be or Not to Be." The Community Health Service became a permanent clinic—the Southeast Health Care Center—in 1976. See "New Health Center for Hunters Point."

77. "First Year in the Forefront."

78. Coleman, "Hunters Point-Bayview Health Service."

79. Nelson, *Body and Soul*, 8.

80. "Panthers Give Food to Needy."

81. On the connection between health and civil rights, and the community health clinic movement, see Loyd, *Health Rights Are Civil Rights*.

82. Woods, *Development Arrested*, 3.

83. Woods, *Development Arrested*, 4.

84. Williams, *Politics of Public Housing*, 14.

85. Local history at this time refers to the women active in Bayview-Hunters Point leadership positions as the "Big Five." I've avoided this moniker because it's misleading—there were far more than five women on the boards of organizations and in other leadership positions at this time.

86. For a more detailed account of this trip, see Brahinsky, "Making and Unmaking of Southeast San Francisco."

87. "Hunters Point Delegation Victorious."

88. "Elouise Westbrook on the New Look."

89. Kelley, *Freedom Dreams*.

90. "Robert Josten to Harvey Rose."

91. With the loss of federal support, the redevelopment of Hunters Point stagnated. A 1983 joint planning study by the SFRA and SFHA, which included longtime housing activists Elouise Westbrook and Julia Comer, noted that the 1960s-era plan had only resulted in 1,630 new housing units—570 less than originally planned. See "Hunters Point Hill Study." Moreover, as Rachel Brahinsky notes in her dissertation, "The Making and Unmaking of Southeast San Francisco," the urban renewal project actually displaced Black families from Hunters Point. However, this was likely because funding cuts after 1972 limited the number of housing units that were eventually constructed, even though demolition of war worker housing had already commenced.

92. "Housing at Hunters Point?"

93. "Statement on Base Realignments."

94. "Hunters Point Naval Shipyard: Special Task Force Preliminary Report"; "Statement of Stanley R. Larsen"; "Hunters Point Economic Re-Use Study"; and "Hunters Point Community History."

95. "Mayor Alioto on Hunters Point Shipyard."

96. "Hunters Point Economic Re-Use Study."

3. THE POLITICS OF ENVIRONMENTAL REPAIR

1. "Alfred E. Engel"; "S.F. Waterfront"; and "Hunters Point Toxics." The city attorney won the case against Triple A, and a $9.3 million fine was imposed. At the time, this was the largest criminal environmental fine in California history. In 1995, a state appeals court reduced the fine to $115,000. See "Toxic Dumping Judgement Cut."

2. The Hunters Point Shipyard was officially closed through the Base Realignment and Closure (BRAC) process, a nationwide action to shutter "surplus" military bases that began after the Cold War. See Sorenson, *Shutting Down the Cold War*. Only a few years prior, San Francisco mayor Diane Feinstein had pursued a bid to homeport the USS *Missouri* at Hunters Point. Yet the ship would have arrived with nuclear weapons on board. An alliance of environmental, antiwar, union, and LGBTQ groups (because of the military's homophobic policies) and ultimately Feinstein's more liberal successor, Art Agnos, successfully opposed the homeporting of the *Missouri*. See Reinhold, "Navy Waging a Battle for Port."

3. U.S. Department of the Navy, Base Realignment and Closure, "Final Historical Radiological Assessment"; and Davis, "Fallout."

4. In March 1993, the Hunters Point Shipyard Citizens Advisory Committee (CAC) approved a framework to guide toxic cleanup at the shipyard: cleanup should be used as a job training-opportunity for local residents; new development should be compatible with tenants already at the shipyard, including artists; job creation should focus on multimedia and biotechnology; and "future

land uses should include acknowledgement of American Indians and blacks who worked at the shipyard in its heyday." See King, "Panel Suggests Guidelines."

5. Selna, "Lennar Corp Dominates Redevelopment"; and Paddock, "Vision for Transforming Hunters Point." According to the journalist Lisa Davis, more than half the cost of building basic infrastructure on the shipyard (such as streets and sewers) would be financed through public bonds. "Arrested Development."

6. Johnson, "Fearful Symmetry of Climate Change."

7. Lennar also secured development rights for two former Marine Air Corps stations in southern California. As of 2023, for various reasons, other companies had won the contracts to develop Mare Island Naval Station, Alameda Naval Air Station, and Concord Naval Weapons Station. Lennar is currently one of several comapnies involved in redeveloping Naval Station Treasure Island.

8. On neoliberal urbanism, see Hackworth, *Neoliberal City*. On some of the specific terms of the deal with Lennar and its potential to reap enormous profits in Hunters Point, see Davis, "Arrested Development." According to local reporting on this issue, to secure only development rights to the shipyard, Lennar cut a deal with three San Francisco–based (though not Bayview-Hunters Point–based) labor and civil rights organizations. In return for their support, Lennar agreed to build additional affordable housing units at the shipyard and give $37.5 million to neighborhood organizations, in the form of grants, through a Community Benefits Agreement (CBA). Fifteen years later, however, the money remained elusive, stalled by political differences on the CBA's Implementation Committee and Lennar's own project delays. See Green, "Infighting Ties Up Millions"; and Yesko and Jones, "Community Awaits Benefits."

9. Lennar's redevelopment project promised to create permanent jobs for Bayview-Hunters Point residents as well, in part by building community facilities and retail space at the shipyard. However, these sources of future jobs were not included in the first phase of development (on Parcel A), which focused on market-rate housing instead. According to Lisa Davis, writing in 2006, "Most of the stated community benefits of the project—public facilities, over 300,000 square feet of retail space, more than a third of the housing designated as affordable, and a significant stretch of open space—are in Parcel B, which, under the Navy's present cleanup schedule, will not be ready for development until sometime after 2008." "Arrested Development." As of this book's publication, Parcel B is still undergoing remediation.

10. Indeed, after Governor Brown's order went into effect, the SFRA's Bayview-Hunters Point Redevelopment project stalled, and it remains an "inactive" project.

11. Jackie is a pseudonym.

12. For other, more theoretical approaches to reparative environmental justice, see Almassi, *Reparative Environmental Justice*; and Papadopoulos, "Chemicals, Ecology, and Reparative Justice."

13. Cram, *Unmaking the Bomb.*

14. "Federal Post-Closure Regulations."

15. Bay Area Regional Health Inequalities Initiative, "Health Inequalities in the Bay Area"; Katz, "Health Programs in Bayview Hunter's Point"; "San Francisco Healthy Homes Project"; Bay Area Air Quality Management District, *Air Monitoring Data*; and Castleman, Bryant, and Kaulukukui, "Concrete Manufacturers."

16. Congressional supporters of the bill were forced to make several important concessions to pass CERCLA. For one, they agreed to a smaller trust fund (an earlier Senate version of the bill included a $4.1 billion fund over six years, but this was reduced to $1.6 billion). Congress also deleted a victims' compensation provision for individuals injured by chemical accidents. See Hird, *Superfund*; and Colten and Skinner, *Road to Love Canal.*

17. During Reagan's first term, the EPA saw a 21 percent budget reduction from FY 1980 to 1983, a decrease in agency funding for research and development, and a decline in EPA staff by 26 percent. EPA referrals of civil cases to the Justice Department decreased by 69 percent. See Sellers et al., *EPA under Siege*; and Fredrickson et al., "Presidential Assaults on Health Protection."

18. Anne Buford Gorsuch, Reagan's first appointment as EPA administrator (and mother of future Supreme Court justice Neil Gorsuch), also resigned from the agency in 1983 after being cited for contempt of Congress. Previously, however, Gorsuch had signaled that, as EPA administrator, she would not seek renewal of Superfund in 1985. See Sellers et al., *EPA Under Siege*; and Hird, *Superfund.*

19. Hird, *Superfund.*

20. Hird, *Superfund*; and Hurley, "From Factory Town to Junkyard."

21. Harvey, "From Managerialism to Entrepreneurialism."

22. U.S Environmental Protection Agency, "Brownfields." In theory, brownfields are less contaminated than Superfund sites. The EPA relies on the numerical Hazardous Ranking System to classify a Superfund site, whereas any plot of land with potential contamination can qualify as a brownfield. Still, the boundary line between a Superfund site and a brownfield can be a politically negotiated, rather than strictly technical, determination. On a tour of another Bay Area military base undergoing remediation and redevelopment in 2018, a retired state environmental regulatory agency staffer took me aside and said, quietly, that the navy had negotiated with the state agencies to keep the base from being declared a Superfund site (which would have subjected the navy to more regulation and likely increased the costs of the base remediation project).

23. Hollifield, "Neoliberalism and Environmental Justice"; Hurley, "From Factory Town to Junkyard"; and Bjelland and Noyes, "Urban Revitalization."

24. Simons, *Turning Brownfields into Greenbacks*, 2.

25. Browner, "Brownfields Becoming Places of Opportunity."

26. Hula, "Changing Priorities and Programs," 182. See also Hourcle and Guenther, "Institutional Controls for Land Use."

27. Bearden, "Comprehensive Environmental Response"; and Hula, "Changing Priorities and Programs."

28. Risk assessment, in an environmental health policy context, is "a process for calculating the probability of certain malign effects . . . caused by exposure to the chemical (or radiological) agent of concern." Crucial to risk assessment calculations are potential pathways by which people can be exposed to specific contaminants—by inhalation, skin contact, or ingestion, for example—and the length of time a person is expected to be exposed to those contaminants. See Applegate, "Risk Assessment, Redevelopment," 254. On risk in environmental policy, see Demortain, *Science of Bureaucracy*; Jasanoff, "Songlines of Risk"; and Nash, "From Safety to Risk."

29. Rocco and Wilson, "Risk-Based Corrective Action"; and Accame and Simon, "Developments in Cleanup Technologies."

30. Applegate, "Risk Assessment, Redevelopment."

31. Krupar, *Hotspotters Report.*

32. The term *background* refers to "an average or expected amount of a substance or radioactive material in a specific environment, or typical amounts of substances that occur naturally in the environment." Agency for Toxic Substances and Disease Registry, "Glossary of Terms—Background." Background levels of a hazardous material are those that would theoretically exist independently of a specific industrial source.

33. Cram, "Unmaking the Bomb," 4. Used in a policy or regulatory context, *risk* is not simply a "danger" or "hazard" (these being more colloquial uses of the term) but refers to a discourse and set of practices for managing uncertainty that emerged first within the Department of Defense in the 1940s and over several decades migrated into federal policymaking, including environmental policy. Risk assessment represents a theoretical approach to decision-making that emphasizes quantification, breaking a problem into discrete components or steps, and the separation of ostensibly "objective" and "subjective" analyses—often depicted as a contrast between "facts" and "values" or between "science" and "politics." One of my arguments in this chapter is that Bayview-Hunters Point residents challenged this distinction between science and politics and insisted on a qualitatively different kind of environmental remediation—more akin to a project of socioecological repair. See Jas and Boudia, *Toxicants, Health and Regulation*; Nash, "From Safety to Risk"; and Demortain, *Science of Bureaucracy.* On "the politics of impossibility" in remediation science, see Cram, *Unmaking the Bomb.* There are other ways of responding to the impossibility of returning landscapes and ecologies to "pristine" conditions that refuse "technoscientific habits" such as risk assessment. See Murphy, "Alterlife."

34. Washburn and Edelmann, "Development of Risk-Based Remediation Strategies."

35. The eleven parcels are Parcels A, B, C, D-1, D-2, E, E-2, F, UC-1, UC-2, and UC-3. The shipyard was initially divided into six parcels.

36. Hunters Point Restoration Advisory Board, Transcript of Meeting, May 27, 2004. I am grateful to Dr. Sumchai for sending me this transcript and highlighting the RAB's concerns with the transfer of Parcel A. They are particularly relevant because in 2018 a radioactive object was discovered near Parcel A, calling into question the whole edifice of remediation, including the promise of state regulatory oversight. See Fagone and Dizikes, "Radioactive Object Found."

37. U.S. Department of the Navy, Base Realignment and Closure, "Draft Decision for Parcel E-2."

38. Policy scholars have expressed concerns about the increased use of institutional controls in remediation projects. See Meyer, "Brownfields and Local Communities"; Hourcle and Guenther, "Institutional Controls"; Wernstedt, Hersh, and Probst, "Basing Superfund Cleanups on Future Uses"; and Wernstedt and Hersh, "'Through a Lens Darkly.'"

39. U.S. Department of the Navy, Base Realignment and Closure, "Record of Decision for Parcel E-2".

40. See *California Military Base Reuse*; and U.S. Environmental Protection Agency, "Superfund Cleanup Process."

41. Andrew is a pseudonym.

42. Jeffrey is a pseudonym.

43. San Francisco Redevelopment Agency, "Summary Regarding Environmental Remediation of Shipyard." As another example, the Hunters Point Shipyard Risk Management Plan, which outlines procedures that must be followed during redevelopment, contains a twenty-eight-page "Unexpected Conditions Response Plan," describing what construction workers should do if they encounter hazardous waste on the shipyard not previously identified by navy remediation documents. San Francisco Civil Grand Jury, "Buried Problems and a Buried Process."

44. Greenaction's map was part of a larger effort to document and address environmental health problems in seven low-income neighborhoods in California: Identifying Violations Affecting Neighborhoods (IVAN), coordinated by Comite Civico del Valle, an environmental justice organization from Imperial County. Each of the seven IVAN projects included a monthly task force meeting that brought representatives from state and municipal agencies together with local residents to discuss and problem-solve reported issues. See Jatkar and London, "From Testimony to Transformation."

45. Greenaction organizers had become concerned about the effects of sea level rise on hazardous waste that would be left on the shipyard after the remediation project was complete. In 2012, Hurricane Sandy hit the New York and New Jersey shorelines, releasing toxic chemicals from dozens of Superfund sites, including sites that had undergone remediation. During the Hunters Point Shipyard's use as a military base and radiological laboratory, the navy introduced an

array of hazardous materials into the soil and groundwater across the shipyard, including substances later banned by the EPA in the 1970s, such as PCBs, DDT, and asbestos. Radioactive isotopes left across the military base included radium-226, cobalt-60, cesium-137, and plutonium-239. Sea level rise threatened to recirculate this archive of military industrialization. See Russell, "Superfund and Climate Change"; and Greenaction for Health and Environmental Justice, *Ticking Time Bomb*.

46. Gena and Kay are Pseudonyms.

47. By *action levels*, Kay referred to the presence of contamination above certain risk levels, which would lead to *remedial actions* (hence, "action levels"). See U.S. Environmental Protection Agency, "Calculating Preliminary Remediation Goals."

48. For other studies of the politics of risk in environmental remediation, see Cram, "Becoming Jane"; and Krupar, *Hotspotters Report*.

49. Cram, "Becoming Jane"; and Cram, *Unmaking the Bomb*.

50. Murphy, "Alterlife and Decolonial Chemical Relations," 495–497. See also Liboiron, Tironi, and Calvillo, "Toxic Politics"; and Ahmann and Kenner, "Breathing Late Industrialism."

51. Liboiron, Tironi, and Calvillo write, "Toxicity is not wayward particles behaving badly. It is not harm at the cellular level. Toxicity includes these things, but we want to avoid fetishizing and reifying the polluting molecule and particle as the locus of toxicity and toxic harm. Structures define toxicity. . . . [T]oxic harm also *maintains* systems, including those that produce inequalities and sacrifice." "Toxic Politics," 333.

52. U.S. Environmental Protection Agency, "RAB Implementation Guidelines." For a comparative analysis of the Hunters Point Shipyard and Monterey Bay Ford Ord RABs, both of which were dissolved by the navy, see Ohayon, "Addressing Environmental Risks?"

53. Sumchai, "RABblerousers!"

54. Welsome, *Plutonium Files*.

55. Massey, "Geographies of Responsibility." One of the more astounding releases of hazardous materials by the navy in Hunters Point was the burning of plutonium-laced fuel oil from Operation Crossroads ships in the shipyard's boilers in 1947. See U.S. Department of the Navy, Base Realignment and Closure, "Final Historical Radiological Assessment."

56. Percy is a pseudonym.

57. See Agency for Toxic Substances and Disease Registry, *Health Consultation: Parcel E Landfill Fire*. Air samples taken on September 1, 2000, showed elevated levels of benzine. Zoellner, "Shipyard Fire Poses No Risk."

58. Sandlos and Keeling, "Zombie Mines and the (Over)burden of History."

59. Dillon, "Pandemonium on the Bay."

60. "Washington Notes."

61. Vasudevan and Smith, "Domestic Geopolitics of Racial Capitalism," 1161–1162.

62. Hess defines *undone science* as "the systemic absence of research identified by counterpublics when they seek to document the potential risks and uncertainties of technologies and industrial processes, and they find that the desired research has not been done or has been significantly underfunded." *Undone Science*, 2. See also Proctor and Schiebinger, *Agnotology*.

63. On miliary environmental remediation and absurdist political tactics, see Krupar, *Hotspotters Report*.

64. Doshi, "Embodied Urban Political Ecology."

65. Hunters Point Restoration Advisory Board, Transcript of Meeting, September 29, 2001, 9.

66. Hunters Point Restoration Advisory Board, Transcript of Meeting, September 29, 2001, 13.

67. Smiley, "Man Who Cried Dust."

68. Wong, "Roots of Black Lives Matter." Ethnic Studies scholar Nicholas Baham described SLAM meetings: "A diverse host of activist liberation theologians and community organizers gather every Thursday evening to hear experts talk about the environmental impact of toxic soil bombardment and asbestos friables released in the development project on the 100-acre Parcel A at the Hunters Point Shipyard." *Coltrane Church*, 195.

69. Schwartz, "Proposition F."

70. Robertson, "New Census Data"; and Wong, "Roots of Black Lives Matter."

71. Arrieta, "Lawsuit, 1,000 Emails."

72. U.S. Department of the Navy, Base Realignment and Closure, "Proposal to Dissolve the Hunters Point Restoration Advisory Board."

73. Ohayon, "Addressing Environmental Risks?"

74. As of June 2020, due to the tireless work of the Bayview-Hunters Point Community Advocates, these transcripts are available online at the UCSF Chemical Industry Documents archive. See www.industrydocuments.ucsf.edu /chemical/.

75. Amara is a pseudonym.

76. Evelyn is a pseudonym.

77. Donna is a pseudonym.

78. Donna could also be understood to ask for what feminist science studies scholar Sandra Harding calls "strong objectivity." Harding critiques the dominant notion of scientific objectivity, which equates clarity and knowledge with value-free science and social distance from the object of study. Harding points out that cultural assumptions and power asymmetries are smuggled into scientific knowledge production at every turn, from which research questions are posed (or not), to how studies are designed, and what counts as scientific evidence. Because it lacks the analytical tools to reflect on these power relations, dominant, or "weak," objectivity reinforces the sociopolitical status quo, including white supremacy. In contrast, strong objectivity begins from the grounded perspective, or standpoint, of marginalized lives, and aims to transform asymmetrical power

relations in knowledge production—such as those that placed the white scientists behind a table facing the largely Black audience of Bayview-Hunters Point residents. Harding, "Strong Objectivity."

79. For example, studies by the nonprofit Committee to Bridge the Gap (CBG) concluded that the military used obsolete safety standards in setting its remedial goals in Hunters Point. Even a University of California, Berkeley, nuclear physicist who disputed CBG's findings told the journalist: "I would be somewhat concerned as a parent moving into [Lennar's] new development," suggesting that even he, the scientist, would not rely on state environmental standards to keep his family safe. See Hirsh et al., "Hunters Point Shipyard Cleanup"; and Roberts, "SF Housing Site Planned." As of this book's publication, environmental nonprofits have continued to raise concerns about the navy's cleanup standards. See Wolfrom, "Fresh Concerns Raised."

80. Almassi, *Reparative Environmental Justice*, 46.

81. Cram, "Unmaking the Bomb," 15.

4. THE DUST OF REDEVELOPMENT

1. The SFHA's name for these units was Hunters View, but most Bayview-Hunters Point residents I met referred to them as West Point, and I use their term here. Angie is a pseudonym.

2. Knight, "Hunters View City's Worst Complex." See also Hunters View Associates, "Hunters View Project Description."

3. Goetz, *New Deal Ruins*. Specifically, the project description for Hunters View proposed 741 units (to replace West Point's 267 units), with 341 units for sale at market rate prices, 267 units classified as public housing, and the rest subsidized at different rates. See Hunters View Associates, LLC, "Hunters View Project Description."

4. Harrison, "Taking of Bayview Hunters Point."

5. De Brito, "Right of Return," 59.

6. Hunters View Community Partners, "Design for Redevelopment."

7. Anderson, Thundiyil, and Stolbach. "Clearing the Air."

8. Mansfield, "Particulate Matters," 1,210.

9. Environmental Law and Justice Clinic, "Letter to the City of San Francisco." See also Katz, "Health Programs in Bayview"; and Rechtschaffen, "Fighting Back against a Power Plant."

10. Mansfield, "Particulate Matters"; and Bell, Zanobetti, and Domimnici, "Evidence on Vulnerability."

11. An exception is Quastel, "Political Ecologies of Gentrification," but he focuses on discourses of nature in gentrification projects, not the material ecologies of redevelopment.

12. Fennell, "Are We All Flint?" On reanimated waste, see also Gregson, Watkins, and Calestani, "Inextinguishable Fibres." On the racial politics of "unbuilding a city," see Safransky, *City after Property*.

13. Noterman, "Fugitive Dust," 858.

14. Kloc, *Air Pollution & Environmental Inequity*.

15. Bayview-Hunters Point residents already experienced higher rates of respiratory illnesses, especially asthma, relative to the rest of the San Francisco Bay Area. See Katz, "Health Programs in Bayview"; and Rechtschaffen, "Fighting Back against a Power Plant."

16. Miriam Solis notes that the census tracts abutting the Southeast Sewage Treatment Plant were 88 percent and 70 percent Black at the time. See Solis, "Conditions and Consequences of ELULU."

17. "Businessmen, Residents Say No Cesspool."

18. Solis, "Conditions and Consequences of ELULU."

19. Specifically, the study, which used air quality data collected by the San Francisco Department of the Environment in 2007, found an up to 900 in one million risk of cancer. At the time, the southeast treatment plant had forty-five diesel backup generators. Kloc, "Air Pollution & Environmental Inequity." Aron is a pseudonym.

20. In 2016, I helped organize a different toxic tour of Bayview-Hunters Point, with Jonathon London (UC Davis) and in collaboration with Greenaction, for attendees of that year's American Association of Geographers conference. Marie Harrison led the tour, and the Southeast Sewage Facility was our first stop.

21. Solis, "Engineering Justice," 41.

22. The community center was remodeled in 2022 along with the renovation of the sewage treatment plant.

23. On landscapes as archives of social movement activism, see Herrera, *Cartographic Memory*.

24. Harrison, "BVHP Mothers Fight." The Mothers Committee, with the Huntersview Tenants Association and Greenaction for Health & Environmental Justice, also produced the report, *Pollution, Health, Environmental Racism, and Injustice: A Toxic Inventory of Bayview Hunters Point, San Francisco*. According to one resident, there was "a coalescing around the environmental justice movement in the 90s and that's when a lot of leaders emerged and bases were built." Solis, "Engineering Justice," 5.

25. Cited in Rechschaffen, "Fighting Back against a Power Plant," 419.

26. Cited in Rechtschaffen, "Fighting Back against a Power Plant," 416.

27. Shapiro, "Attuning to the Chemosphere."

28. The study specifically looked at the years 1988–1992. See Rubenstein, "S.F. Rally against Cancer"; Jane Kay, "Pollution Fears Stir Activists"; and Rechtschaffen, "Fighting Back against a Power Plant."

29. "Hunters Point Power Plant Reaches Impasse"; Brummer-Kocks and Richardson, "Help Clean Up Bayview"; and Ramo, "Hunters Point."

30. The other context for this decision was energy deregulation; Governor Jerry Brown deregulated the state energy industry in 1996. Personal communication with Anne Eng. See also Epstein, "Advocates Push S.F. to Buy Power Plants."

31. Harrison, "Air Up Here."

32. At this time, the company also applied for emergency approval to locate an additional power plant on a floating barge offshore southeast San Francisco to provide "standby generation and voltage support" during high peak energy use (the governor's emergency approval would have allowed PG&E to use the barge without adhering to the California Environmental Quality Act). The 90-by-300-foot offshore barge, with four five-story jet-fueled turbines, was originally designated for Hunters Point. The company later shifted the proposed location of the barge to the San Francisco International Airport, a few miles south of Bayview-Hunters Point, in response to strong neighborhood opposition. The barge, which had been en route to San Francisco through the Panama Canal, was reported on by the *San Francisco Chronicle* and ultimately rejected by city officials. See Millard, "Power Plant a 'Fantasy'"; Kay, "Activists Vow to Stop Power Ship"; and California Independent Systems Operator, "Memorandum."

33. Goodyear, "Protest against PGE Plant"; and "Best of the Bay 2006."

34. "Victory for Hunters Point Activist."

35. Romans, "Learn about Serpentinite"; and U.S. EPA, "Naturally Occurring Asbestos."

36. According to an evaluation later conducted by the Agency for Toxic Substances and Disease Registry (ATSDR), "There are no asbestos monitoring data available for the first few months of grading (April 25, 2006–August 2, 2006) due to operator error and equipment malfunctions." Agency for Toxic Substances and Disease Registry, *Health Consultation: Parcel E Landfill Fire.*

37. The ATSDR report details these air quality exceedences and lag times. In one example from the report: "Between August 3 and August 10, 2006, asbestos levels exceeded 16,000 structures/m3 [the allowable concentration or threshold of asbestos] on three days (no measurement reported on three of the seven days), with a maximum level of asbestos measured at 24,400 structures/m3." Despite this hazardous air quality, "the asbestos results from the beginning of August were not received until August 14." In other words, asbestos levels exceeded allowable concentrations of 16,000 structures/m3 on three out of the four days that monitoring took place, yet Lennar's construction activities did not stop until eleven days after the first of these air quality exceedences. Agency for Toxic Substances and Disease Registry, *Health Consultation: Parcel E Landfill Fire.*

38. From the report: "According to the manufacturer, the instrument that has been used to monitor dust at Parcel A is designed for personal/breathing zone monitoring, plant walk-through surveys, remediation site worker exposure monitoring, and indoor air quality. The instrument being used is sensitive

to moisture and is a passive sampler. Dust monitors that are approved for PM 10 ambient air standards by the California Air Resources Board are all active samplers. Further, there are dust monitors available that are designed for outdoor applications where moisture is present. Due to the novel application of the equipment for fence line monitoring, CDPH [California Department of Public Health, the agency that conducted the report for ATSDR] is not able to interpret whether dust exposures in the community occurred that would explain some of the respiratory symptoms, nausea, and vomiting. We recommend using dust monitors that have been certified for fence line monitoring." Agency for Toxic Substances and Disease Registry, *Health Consultation: Parcel E Landfill Fire*.

39. Smiley, "Man Who Cried Dust."

40. San Francisco Board of Education Resolution 79-25A1 (2006).

41. The report was prepared by the California Department of Public Health for the Agency Toxic Substances Disease Registry, through a cooperative agreement between the two agencies.

42. "San Francisco Healthy Homes Project."

43. Phelan, "Air District Fined Lennar Half a Million." Additionally, in 2007 two Lennar employees filed a lawsuit against the company, alleging retaliation for voicing concerns about asbestos dust and that the company's monitoring equipment wasn't functioning properly. The employees later settled with Lennar.

44. Murphy, *Sick Building Syndrome*. Indeed—to my mind—both activists and health professionals likely overemphasized the issue of asbestos, though for different reasons. Asbestos is a known carcinogen and regulated by federal law. In contrast, construction dust is difficult to classify, and more easily dismissed as an irritant rather than a health hazard. It seems understandable that activists would focus on asbestos precisely because it was more likely to garner attention from the state and news media. As a separate point, if public health experts thought residents were wrong in drawing connections between their health problems and Lennar's construction dust, it is worth considering how, in a place where people are sick and yet proof of toxic exposure remains elusive, "Paranoia is a perfectly reasonable response." Ahmann, "Uncertainty in Motion," 314. This is not to say that residents' reported health symptoms were *not* connected to construction dust, only that dismissing activists' focus on asbestos in particular misses the bigger picture of how power and vulnerability operate in "late industrial" places.

45. McCormick, "Southeast Side Residents."

46. Five Point, "Candlestick Point,"; and San Francisco Office of Community Investment and Infrastructure, "Candlestick Point."

47. Matier & Ross. "Candlestick Will Go Out"; and Bowles, "Can Developers Turn a Neglected Neighborhood?"

48. In the words of the addendum to the EIR, "the demolition by implosion would result in no new significant impact." San Francisco Planning Department,

Addendum 3, 15. Bayview Hill is different from "the hill" in Hunters Point, referenced in chapter 2.

49. Alice is a pseudonym.

50. Golden Gate University's Environmental Law and Justice Clinic sent a letter to the city of San Francisco regarding the stadium implosion on November 18, 2014. The authors of the letter wrote: "The Addendum fails to identify and analyze in an adequate manner the impacts of imploding Candlestick Park stadium." Further, "The addendum fails to identify the amount of concrete and other material that would be imploded; fails to characterize the kind and amount of PM matter the implosion would generate; [and] the extent to which different kind of PM would disperse under wind conditions relevant to where the stadium is located." Environmental Law and Justice Clinic, Golden Gate School of Law, "Letter to the City of San Francisco," 6.

51. Jenny is a pseudonym.

52. One of the nicknames for Candlestick Stadium was "Cave of the Winds." Or, as an essayist explained in an ode to the stadium on the eve of its teardown, grievances with Candlestick were numerous and included "the hat-ripping gusts of wind that have for decades bedeviled quarterbacks and made utter mockery of the science of field goal kicking." Hosseini, "Closing the Cave of the Winds." Even the company contracted to carry out the stadium implosion recognized that wind is "robust . . . at and around the stadium." Quoted in Environmental Law and Justice Clinic, Golden Gate School of Law, "Letter to the City of San Francisco," 6.

53. San Francisco Planning Department, *Addendum 3*, 14.

54. Troy is a pseudonym.

55. For example, reports published in 2017 and 2020 by the Golden Gate University School of Law's Environmental Law and Justice Clinic show how regulatory neglect in Bayview-Hunters Point is a systemic issue. The reports detail how the Bay Area Air Quality Management District's practices allow concrete manufacturers in the neighborhood to operate without permits, and they document significant delays in both permitting decisions and enforcement of permit violations. See Castleman, Bryant, and Kaulukukui, "Concrete Manufacturers and Regulatory Role"; and Santos et al., "Concrete Production and Regulatory Role."

56. Pezzullo, *Toxic Tourism*, 9–10.

57. Murphy, *Sick Building Syndrome*, 9.

58. Jeffries, "Force Us to Face Anti-Blackness."

59. Dillon, "Breathers of Bayview Hill."

60. Sharpe, *In the Wake*.

61. Angelo, *How Green Became Good*, 21–23.

62. Angelo, *How Green Became Good*, 174.

63. On urban greening as a well-intentioned public good, see Angelo, *How Green Became Good*.

64. On the U.S. military presenting itself as environmental stewards, see also Krupar, *Hotspotters Report*.

65. San Francisco Office of Community Investment and Infrastructure, "Hunters Point Shipyard/Candlestick Point," 29.

66. Krupar, *Hotspotters Report*, 2.

67. In writing these, I am thinking of Shiloh Krupar, who writes, "Attendant to the establishment of the RMANWR [Rocky Mountain Arsenal National Wildlife Refuge], activities have multiplied for the purposes of public relations and image management, real-estate development, pedagogical investments, and military restructuring." *Hotspotter's Report*, 64.

68. "Lennar Loss Narrower Than Expected."

69. Lennar, "Hunters Point Shipyard" (June 24, 2015).

70. Greenberg, "'Sustainability Edge.'"

71. Greenberg, "'Sustainability Edge,'" 118.

72. Lochhead, "EB-5 Visas Given to Investors."

73. Selling nature and sustainability represented a pivot from earlier advertising directed to the Bayview-Hunters Point community. Lennar's websites from before 2008 directly addressed concerns held by Bayview-Hunters Point residents about what redevelopment at the shipyard would mean for the longtime community and who would benefit from the project. An early Hunters Point Shipyard website by Lennar, in 2005, for example, featured the color orange and emphasized, at the top of its home page, "jobs, homes, and opportunity in our community." The phrase "our community" positioned Lennar as working for and with current Bayview-Hunters Point residents, as if redevelopment were a community-based endeavor and not part of the company's portfolio of military base redevelopment projects. A redesigned website later that same year relied on light gray as its main color, with muted blue, green, and purple accents. The website's home page advertised a Homebuyer's Assistance and Community Benefits program. "The Renaissance of the Hunters Point Shipyard," displayed as a banner across the top of the home page, served as Lennar's slogan, again promoting the shipyard as a form of economic development for current residents and implicitly addressing their history of social marginalization: "This is a historic moment for the Bayview-Hunters Point community. By working together, we are helping the Shipyard and the community to enjoy the renaissance they deserve." Images of the shipyard's future on the company's website at this time depict a visibly attractive landscape; computer-generated aerial views show most of the shipyard as green "open space." Still, the text and colors of the website design continue to direct attention to business opportunities and assurances that the shipyard would remain connected to the longtime Black community in Bayview-Hunters Point. See Lennar, "Hunters Point Shipyard" (March 10, 2005) and Lennar, "Hunters Point Shipyard" (November 25, 2005).

74. Ross, "Hunters Point Housing Streamlines." See also, Dineen, "Lennar Taps Mark Co."

75. On Blackness as risk, see Rios, *Black Lives and Spatial Matters*.

76. Taylor, *Race for Profit*, 9.

77. Taylor, *Race for Profit*. See also Bledsoe and Wright, "Anti-Blackness of Global Capital."

78. Browne, *Dark Matters*, 16.

79. Lefebvre, *Production of Space*.

80. Merrifield, "Place and Space," 523.

81. Angela is a pseudonym.

82. Merrifield, "Place and Space," 523.

83. Bruno, "Work of Repair."

84. Summers, *Black in Place*, 3.

85. In a way, this chapter also responds to Doshi's outline of an "embodied political ecology."

86. Ahmann, "Uncertainty in Motion," 306.

87. Greenberg, "What on Earth Is Sustainable?," 58.

CONCLUSION

1. Whiting, "S.F. About to Break Ground."

2. To learn more about Adam Rogers, see Agee, *Streets of San Francisco*.

3. June is a pseudonym.

4. McKittrick, *Demonic Grounds*, ix.

5. This book has touched on major environmental justice campaigns and focused on remediation and redevelopment at the shipyard. But it is not an exhaustive study of the landscape of environmental justice work, broadly defined, in Bayview-Hunters Point. One important new project is the Hunters Point Biomonitoring Project, led by Dr. Ahimsa Porter Sumchai. See Sumchai, "HP Biomonitoring Project."

6. In this, they reveal what Melissa Checker calls "the paradoxes facing environmental justice activists who fight for healthier neighborhoods that are also affordable." *Sustainability Myth*, 7.

7. Molanphy, "'We Feel Abandoned.'"

8. "AB3121."

9. *California Reparations Report*. For some reflections on the work of the California Reparations Task Force, see Lewis, "Black Life Beyond Injury."

10. Kelley, *Freedom Dreams*, 129.

11. Movement for Black Lives, "Reparations Now! Toolkit"; and Hartman, *Lose Your Mother*.

12. Coates, "Case for Reparations."

13. Indeed, Bayview-Hunters Point residents who rent or own single-family homes in the neighborhood—outside of the city's public housing projects—were disproportionately impacted by the foreclosure crisis in 2008. Between 2008 and 2012, an estimated fifteen hundred households (or 15 percent of the neighborhood's housing stock) in the Bayview-Hunters Point zip code were foreclosed on by their lending banks. Samaha, "Dispossessed."

14. The city of Evanston, Illinois, for example, recently began to distribute what it calls "local reparations," starting with a housing program aimed at addressing the effects of redlining and the historical racial wealth gap in the city. The program has recieved mixed and in some cases quite critical reception by local residents. See Misra, "Illinois City's Reparations Plan."

15. San Francisco African American Reparations Advisory Committee, *Efforts to Support Reparations Plan*, 14.

16. The Task Force worked with William Darity and Kirsten Mullen, two economists who have written extensively on reparations. See Darity and Mullen, *From Here to Equality*.

17. Ranganathan, "Thinking with Flint," 22.

18. Táíwò, *Reconsidering Reparations*, 74.

19. Táíwò, *Reconsidering Reparations*, 72.

20. Táíwò also compares the constructive view of reparations with Cold War–era anticolonial national independence movements, which sought to build a new global political and economic order apart from U.S. capitalism and U.S.S.R. state socialism. This new political and economic order would be constructed through "the recognizable move of reparations politics: redistribution of global wealth, from the First World (back) to the Third World." Táíwò notes that, at the national level, the 1960s-era, Detroit-based Republic of New Afrika (RNA) made similar claims, demanding the U.S. government cede land and provide funds to establish a separate Black nation. Reparations, for the RNA, concerned the resources (land and money) necessary to "assure the new nation a reasonable chance of success"; in short, to support Black self-determination. Táíwò, *Reconsidering Reparations*, 71, quoting Dr. Nkechi Taifa.

21. Costley, "Defunding the Police." Indeed, Táíwò argues that climate justice must be a central part of reparations for global racial empire, as the "climate crisis arises from the same political history as racial injustice and presents a challenge of the same scale and scope. The transformations we succeed or fail to make in the face of the climate crisis will be decisive for the project of racial justice, and vice versa." Táíwò, *Reconsidering Reparations*, 147.

22. According to geographer Jovan Scott Lewis, we are living in a "reparative conjuncture" in which "society has begun to reckon with, if not reconcile, the fact that our world was made possible by a wide-ranging set of ongoing injuries." Scott, who served on the California Reparations Task Force, distinguishes between "reparations" and "repair" and argues that "to fully conceive of constitutes

Black repair, we must fully disaggregate Black harm and the practice of world-making. By doing so, we, at the very least, discursively liberate Black being from suffering." This book is focused on racism and resistance in San Francisco, and does not seriously engage with Scott's call to conceive of "Black life beyond injury" (or, similarly, as Katherine McKittrick puts it, how "racial violences shape . . . but do not wholly define, Black worlds"). Nor is my conclusion infomed by Lewis's recent article, which was published as I received my book proofs. I want to mark that absence, while still making a case that this book can contribute to conversations about the ways our 'reparative conjuncture" can reshape environmental justice theory and politics. See Lewis, "Black Life Beyond Repair," 1–2, 5.

23. California Task Force to Study and Develop Reparations Proposals for African Americans, *Final Report*, 48.

24. Ranganathan, "Thinking with Flint," 29. Or, as Best and Hartman, quoting Farley, write, "the state cannot grant the prayer for reparations; it cannot without destroying itself." Best and Hartman, "Fugitive Justice," 3.

25. Kelley, *Freedom Dreams*, 10.

Bibliography

AB3121: Task Force to Study and Develop Reparations Proposals for African Americans. Office of the Attorney General, State of California. Accessed July 8, 2023, https://oag.ca.gov/ab3121.

Accame, Guillermo M., and John A. Simon. "Recent Developments in Cleanup Technologies." *Remediation Journal* 9, no. 1 (1998): 123–132.

Agee, Christopher. *The Streets of San Francisco: Policing and the Creation of a Cosmopolitan Liberal Politics, 1950–1972.* Chicago: University of Chicago Press, 2014.

Agency for Toxic Substances and Disease Registry. "Glossary of Terms— Background." Accessed October 12, 2023. www.atsdr.cdc.gov/glossary.html.

———. *Health Consultation: Parcel E Landfill Fire at Hunters Point Shipyard.* March 2, 2001. Printout in author's possession.

———. *Health Consultation: Review of Asbestos Air Monitoring Data Collected During Grading Activities at the Hunter's Point Parcel A.* September 30, 2008. www.atsdr.cdc.gov/HAC/pha/ParcelABayviewHuntersPoint/Parcel%20A_Bayview_Hunters_%20Point%20HC%209-30-2008.pdf.

Ahmann, Chloe. "Uncertainty in Motion: Rumors of a Proxy War in Late Industrial Baltimore." *Cultural Anthropology* 38, no. 3 (2023): 303–333.

Ahmann, Chloe, and Alison Kenner. "Breathing Late Industrialism." *Engaging Science, Technology, and Society* 6 (2020): 416–438. "Alfred E. Engel—Head of S.F.'s Triple A Shipyard." *San Francisco Chronicle*, May 28, 1987.

Akins, Damon B., and William J. Bauer. *We Are the Land: A History of Native California*. University of California Press, 2021.

"Alice Griffith Projects." *Spokesman*, September 2, 1965.

Allen, Robert. *Port Chicago Mutiny: The Story of the Largest Mass Mutiny Trial in US Naval History*. Heyday, 1989.

Allenworth, Ariana Faye, Adrienne Hall, and Erin McElroy. "(Dis)location/ Black Exodus and the Anti-Eviction Mapping Project." The Abusable Past, August 6, 2019. www.radicalhistoryreview.org/abusablepast/dislocation -black-exodus-and-the-anti-eviction-mapping-project/.

Almaguer, Tomas. *Racial Fault Lines: The Historical Origins of White Supremacy in California*. University of California Press, 1994.

Almassi, Ben. *Reparative Environmental Justice in a World of Wounds*. Lexington Books, 2020."Alioto Pledge on Hunters Point." Joseph L. Alioto Papers, 1958–1977, folder 11, box 9. San Francisco Public Library, San Francisco.

Anderson, M. Kat. *Tending the Wild: Native American Knowledge and the Management of California's Natural Resources*. University of California Press, 2004.

Anderson, M. Kat, Michael G. Barbour, and Valerie Whitworth. "A World of Balance and Plenty: Land, Plants, Animals, and Humans in a Pre-European California." *California History* 76, nos. 2/3 (1997): 12–47.

Anderson, Jonathan O., Josef G. Thundiyil, and Andrew Stolbach. "Clearing the Air: A Review of the Effects of Particulate Matter Air Pollution on Human Health." *Journal of Medical Toxicology* 8, no. 2 (2012): 166–175.

Andrews, Richard N. L. *Managing the Environment, Managing Ourselves: A History of American Environmental Policy*. Yale University Press, 2006.

Angelo, Hillary. *How Green Became Good: Urbanized Nature and the Making of Cities and Citizens*. University of Chicago Press, 2021.

Anguelovski, Isabelle. "From Toxic Sites to Parks as (Green) LULUs? New Challenges of Inequity, Privilege, Gentrification, and Exclusion for Urban Environmental Justice." *Journal of Planning Literature* 31, no. 1 (2016): 23–36.

Anguelovski, Isabelle, Anna Livia Brand, James J. T. Connolly, Esteve Corbera, Panagiota Kotsila, Justin Steil, Melissa Garcia-Lamarca, et al. "Expanding the Boundaries of Justice in Urban Greening Scholarship: Toward an Emancipatory, Antisubordination, Intersectional, and Relational Approach." *Annals of the American Association of Geographers* 110, no. 6 (2020): 1743–1769.

Annual Report, U.S. Naval Radiological Defense Laboratory, 1953. U.S. Department of Energy, Nuclear Testing Archive. Accession number NV0756422. https://www.osti.gov/opennet/detail?osti-id=16005500.

Applegate, John S. "Risk Assessment, Redevelopment, and Environmental Justice: Evaluating the Brownfields Bargain." *Journal of Natural Resources & Environmental Law* 13 (1997): 243.

Arbona, Javier. "After the Blast: Building and Unbuilding Memories of Port Chicago." PhD diss., University of California Berkeley, 2013.

———. "Anti-memorials and World War II Heritage in the San Francisco Bay Area: Spaces of the 1942 Black Sailors' Uprising." *Landscape Journal* 34, no. 2 (2015): 177–192.

Arrieta, Rose. "A Lawsuit, 1,000 Emails, and San Francisco's Superfund Site." *In These Times*, March 24, 2011.

Baham, Nicholas Louis, III. *The Coltrane Church: Apostles of Sound, Agents of Social Justice*. McFarland, 2015.

Baranski, John. *Housing the City by the Bay: Tenant Activism, Civil Rights, and Class Politics in San Francisco*. Stanford University Press, 2019.

Bay Area Air Quality Management District. *Air Monitoring Data: Review of Air Monitoring Data for Bayview Hunters Point*. August 2022. www.baaqmd .gov/~/media/files/ab617-community-health/bayview-hunters-point /documents/bvhp_monitoring_overview_force_20220818-pdf.pdf?la=en& rev=e291d02f7e4c42618d7361bf2b0ebfd1.

Bay Area Regional Health Inequities Initiative. "Health Inequalities in the Bay Area." Accessed March 3, 2022, https://bd74492d-1deb-4c41-8765-52b2 e1753891.filesusr.com/ugd/43f9bc_f79cccc60ed7424faea668cc75d5736f.pdf.

"Bayview an Organized Community." *Spokesman*, January 20, 1966.

"Bayview Community Speak Out at Housing Authority Meeting." TV broadcast, March 9, 1966, KRON-TV. Bay Area TV Archive, San Francisco State University.

"Bayview Is to Have More Homes." *San Francisco Call*, May 25, 1912.

Bayview-Hunters Point Area B Survey, San Francisco, California: Historic Context Statement. Prepared by Kelley & VerPlanck for the San Francisco Redevelopment Agency, February 11, 2010. www.sanfranciscohistory.com /BVHP_Context.pdf.

Bearden, David M. "Comprehensive Environmental Response, Compensation, and Liability Act: A Summary of Superfund Cleanup Authorities and Related Provisions of the Act." *Congressional Research Service*, June 14, 2012.

Becker, Natalie, and Marjorie Myhill. *Power and Participation in the San Francisco Community Action Program, 1964–1967*. Institute of Urban and Regional Development, University of California, Berkeley, 1967. https:// escholarship.org/uc/item/9gf1v6s9.

Beckett, Lois. "Standing Up with the Aboriginal Blackmen United." *SF Weekly*, July 28, 2010.

Bell, Michelle, Antonella Zanobetti, and Francesca Dominici. "Evidence on Vulnerability and Susceptibility to Health Risks Associated with Short-Term Exposure to Particulate Matter: A Systematic Review and Meta-analysis." *American Journal of Epidemiology* 178, no. 6 (2013): 865–876.

Best, Stephen, and Saidiya Hartman. "Fugitive Justice." *Representations* 92, no. 1 (2005): 1–15.

"Best of the Bay 2006—Local Hero." *San Francisco Bay Guardian*, July 26–August 1, 2006.

"Big Victory for Hunters Point Activist." *San Francisco Chronicle*, May 15, 2006.

Bishari, Nuala Sawyer. "Two Imprisoned for Hunters Point Shipyard Cleanup Fiasco." *SF Weekly*, May 4, 2018.

Bjelland, Mark D., and Ian Noyes. "Urban Revitalization in a Neoliberal Key." In *Urban Transformations: Geographies of Renewal and Creative Change*, edited by Nicolas Wise and Julie Clark, 43–61. Taylor & Francis, 2017.

Bledsoe, Adam, and Willie Jamaal Wright. "The Anti-Blackness of Global Capital." *Environment and Planning D: Society and Space* 37, no. 1 (2019): 8–26.

Blomley, Nicholas. *Unsettling the City: Urban Land and the Politics of Property*. Routledge, 2004.

Bradley, David. *No Place to Hide*. Little, Brown, 1948.

Booker, Matthew. *Down by the Bay: San Francisco's History Between the Tides*. University of California Press, 2013.

Bowles, Nellie. "Can Developers Turn a Neglected SF Neighborhood into an Innovation Hub." *Vox*, October 15, 2014. www.vox.com/2014/10/15/11631912 /can-developers-turn-a-neglected-sf-neighborhood-into-an-innovation-hub.

Brahinsky, Rachel. "'Hush Puppies,' Communalist Politics, and Demolition Governance: The Rise and Fall of the Black Fillmore." In *Ten Years That Shook the City, 1968–1978*, edited by Chris Carlsson, 141–153. City Lights Books, 2011.

———. "The Making and Unmaking of Southeast San Francisco." PhD diss., University of California, Berkeley, 2012.

———. "Race and the Making of Southeast San Francisco: Towards a Theory of Race-Class." *Antipode* 46, no. 5 (2014): 1258–1276.

Brechin, Gray. *Imperial San Francisco: Urban Power, Earthly Ruin*. University of California Press, 2006.

Broussard, Albert. *Black San Francisco: The Struggle for Racial Equity in the West, 1900–1954*. University of Kansas Press, 1993.

Brown, Kendall. *A History of Mining in Latin America: From the Colonial Era to the Present*. University of New Mexico Press, 2012.

Browne, Simone. *Dark Matters: On the Surveillance of Blackness*. Duke University Press, 2015.

Browner, Carol M. "Brownfields Are Becoming Places of Opportunity." *Journal of Natural Resources & Environmental Law* 13, no. 2 (1998). https:// uknowledge.uky.edu/cgi/viewcontent.cgi?article=1256&context=jnrel.

Brummer-Kocks, Wendy, and Linda Richardson. "Help Clean Up Bayview." Unidentified, undated newspaper clipping. Archives of Greenaction for Health and Environmental Justice, San Francisco.

Bruno, Tianna. "The Work of Repair, Land Relations, and Pedagogy." Panel presentation, Association of American Geographers Annual Conference, Denver, 2023.

"Businessmen, Residents Say No Cesspool in Hunters Point." *Sun-Reporter*, June 7, 1975.

"BVHP Joint Housing Committee Constitution." Joseph L. Alioto Papers, 1958–1977, box 14, folder 41. San Francisco Public Library, San Francisco.

Cairns, Kate. "Caring about Water in Camden, New Jersey: Social Reproduction against Slow Violence." *Gender, Place & Culture* 29, no. 10 (2022): 1423–1445.

California Independent Systems Operator. "Memorandum." June 28, 2000. Archives of Greenaction for Health and Environmental Justice, San Francisco.

California Military Base Reuse. Department of Toxic Substances Control. State of California, n.d. https://dtsc.ca.gov/wp-content/uploads/sites/31/2017/01/Brochure-no-CL.pdf.

The California Reparations Report. Office of the Attorney General, State of California. Accessed July 8, 2023. https://oag.ca.gov/ab3121/report.

Canter, Donald. "Hunters Point Ownership in Questions." *SF Examiner*, April 16, 1968.

Capozzola, Christopher. "Jacob Lawrence: Historian." *Rethinking History* 10, no. 2 (2006): 291–295.

Castleman, Steve, Amber Bryant, and Kawika Kaulukukui. "Concrete Manufacturers and the Regulatory Role of the Bay Area Air Quality Management District." Golden Gate University School of Law, Environmental Law and Justice Clinic, May 25, 2017.

Caulkins, Robert. *An Economic and Industrial Survey of the San Francisco Bay Area.* Sacramento: California State Planning Board, 1941.

"The Character of Mrs. Essie Webb." *Spokesman*, January 6, 1966.

Checker, Melissa. *The Sustainability Myth: Environmental Gentrification and the Politics of Justice.* NYU Press, 2020.

———. "Wiped Out by the 'Greenwave': Environmental Gentrification and the Paradoxical Politics of Urban Sustainability." *City & Society* 23, no. 2 (2011): 210–229.

Cebul, Brent. "Tearing Down Black America." *Boston Review*, July 22, 2020.

Coates, Ta-Nehisi. "The Case for Reparations." *Atlantic*, June 2014.

Coleman, Arthur. "The Hunters Point-Bayview Community Health Service." *California Medicine*, March 1969.

Colten, Craig E., and Peter N. Skinner. *The Road to Love Canal: Managing Industrial Waste before EPA.* University of Texas Press, 2010.

Commercial and Industrial Activity in San Francisco: Present Characteristics and Future Trends. Arthur D. Little, June 1975.

"Cop Goes Free." *Spokesman*, October 29, 1966.

"Core Community Benefits Agreement: Hunters Point Shipyard/Candlestick Point Integrated Development Project." Accessed March 3, 2022. http://d10benefits.org/wp-content/uploads/2021/04/Lennar_AD10-CCBA-Executed59076764_1.pdf.

Correia, David. "Making Destiny Manifest: United States Territorial Expansion and the Dispossession of Two Mexican Property Claims in New Mexico, 1824–1899." *Journal of Historical Geography* 35, no. 1 (2009): 87–103.

Costley, Drew. "Defunding the Police Is an Environmental Justice Issue." *OneZero*. June 17, 2020. https://onezero.medium.com/defunding-the-police-is-an-environmental-justice-issue-9c14e48e1ce5.

Cram, Shannon. "Becoming Jane: The Making and Unmaking of Hanford's Nuclear Body." *Environment and Planning D: Society and Space* 33, no. 5 (2015): 796–812.

———. *Unmaking the Bomb: Environmental Cleanup and the Politics of Impossibility.* University of California Press, 2023.

Crocker and Langley Business Directory. 1901. Bancroft Library, University of California Berkeley, 1901.

Crosby, Alfred W. *Ecological Imperialism: The Biological Expansion of Europe, 900–1900.* Cambridge University Press, 2004.

Crowe, Daniel. *Prophets of Rage: The Black Freedom Struggle in San Francisco, 1945–1969.* Routledge, 2018.

Curran, Winifred, and Trina Hamilton. "Just Green Enough: Contesting Environmental Gentrification in Greenpoint, Brooklyn." *Local Environment* 17, no. 9 (2012): 1027–1042.

Daniels, Douglas Henry. *Pioneer Urbanites: A Social and Cultural History of Black San Francisco.* University of California Press, 1990.

Darity, William A., Jr., and A. Kirsten Mullen. *From Here to Equality: Reparations for Black Americans in the Twenty-First Century.* UNC Press Books, 2022.

Darling, Katherine Weatherford, Jenny Reardon, Andy Murray, Emily Caramelli, Dennis Browe, Nikobi Petronelli, and Emma Mitchell-Sparke. "Just Biomedicine on Third Street? Health and Wealth Inequalities in San Francisco's Biotech Hub." In *Counterpoints: A San Francisco Atlas*, edited by Anti-Eviction Mapping Project, 135–152. PM Press, 2021.

Davis, Horace B. "Nations, Colonies and Social Classes: The Position of Marx and Engels." *Science & Society* 29, no. 1 (1965): 26–43.

Davis, Lisa. "Arrested Development." *SF Weekly*, November 19, 2003.

———. "Fallout: The Past Is Present; The Nuclear Witness." *SF Weekly*, July 31, 2002.

De Brito, Deia. "Right of Return: Public Housing in San Francisco's Hunters Point." *Race, Poverty & the Environment* 16, no. 1 (Spring 2009): 58–60.

"Declaring Policy Affecting Hunters Point War Housing Projects and Development of a Master Plan for Improvement of the Hunters Point Area."

December 13, 1965. John F. "Jack" Shelley Papers, 1953–1967, box 2, folder 20. San Francisco Public Library, San Francisco.

Demortain, David. *The Science of Bureaucracy: Risk Decision-Making and the US Environmental Protection Agency.* MIT Press, 2020.

"Description of Hunters Point." In "A Report to the Economic Opportunity Council of San Francisco on Social and Economic Conditions," November 29, 1965. John F. "Jack" Shelley Papers, 1953–1967, folder 12, box 2. San Francisco Public Library, San Francisco.

Dhillon, Jaskiran. "Indigenous Resistance, Planetary Dystopia, and the Politics of Environmental Justice." *Globalizations* 18, no. 6 (2021): 898–911.

Dillon, Lindsey. "The Breathers of Bayview Hill: Redevelopment and Environmental Justice in Southeast San Francisco." *UC Law Environmental Journal* 24 (2018): 227–236.

———. "Civilizing Swamps in California: Formations of Race, Nature, and Property in the Nineteenth Century US West." *Environment and Planning D: Society and Space* 40, no. 2 (2022): 258–275.

———. "Crossroads in San Francisco." In *Inevitably Toxic: Historical Perspectives on Contamination, Exposure, and Expertise*, edited by Sarathy, Brinda, Vivien Hamilton, and Janet Farrell Brodie, 74–96. University of Pittsburgh Press, 2018.

———. "Pandemonium on the Bay: Naval Station Treasure Island and the Toxic Legacies of Atomic Defense." In *Urban Reinventions: San Francisco's Treasure Island*, edited by Lynne Horiuchi and Tanu Sankalia, 140–158. University of Hawaii Press, 2017.

———. "Race, Waste, and Space: Brownfield Redevelopment and Environmental Justice at the Hunters Point Shipyard." *Antipode* 46, no. 5 (2014): 1205–1221.

———. "The San Francisco Blues." In *The Black Geographic: Praxis, Resistance, Futurity*, edited by Camilla Hawthorne and Jovan Scott Lewis, 246–263. Duke University Press, 2023.

Dillon, Lindsey, and Julie Sze. "Police Power and Particulate Matters: Environmental Justice and the Spatialities of In/Securities in US Cities." *English Language Notes* 54, no. 2 (2016): 13–23.

Dineen, J. K. "Infusion of Housing Delivers Hope to SF, Helps Families." *San Francisco Chronicle*, November 2, 2018.

———. "Lennar Taps the Mark Co. for Shipyard Marketing." *San Francisco Business Times*, January 21, 2014.

Doshi, Sapana. "Embodied Urban Political Ecology: Five Propositions." *Area* 49, no. 1 (2017): 125128.

Dow, Gerald. 1973. "Bay Fill in San Francisco: A History of Change." MA thesis, San Francisco State University.

Du Bois, W. E. B. "Colored California." *Crisis*, August 1913.

Dunn-Salahuddin, Aliyah. "A Forgotten Community, A Forgotten History: San Francisco's 1966 Uprising." In *The Strange Careers of the Jim Crow North:*

Segregation and Struggle Outside of the South, edited by Komozi Woodard and Jeanne Theoharis, 211–234. NYU Press, 2019.

Ebasco Services. *Port of San Francisco, Facilities Improvement Survey for the Port Authority*. New York, 1959.

"Elouise Westbrook on the New Look for Hunters Point." *Eyewitness News*. Aired October 25, 1972, on KPIX. Bay Area TV Archive, San Francisco State University.

Engstrand, Iris H. W. "Seekers of the 'Northern Mystery': European Exploration of California and the Pacific." In *Contested Eden: California before the Gold Rush*, edited by Ramón A. Gutiérrez and Richard J. Orsi, 78–110. University of California Press, 1998.

Environmental Law and Justice Clinic, Golden Gate School of Law, "Letter to the City of San Francisco Concerning Hunters Point Candlestick Park Implosion." November 8, 2014. https://digitalcommons.law.ggu.edu/cgi /viewcontent.cgi?article=1029&context=eljc

Epstein, Edward. "Advocates Push S.F. to Buy Power Plants." *San Francisco Chronicle*, March 9, 2003.

Examination of the War on Poverty: Hearings before the Subcommittee on Employment, Manpower, and Poverty of the Senate Comm. on Labor and Public Welfare. 19th Cong. (May 19, 1967). https://babel.hathitrust.org/cgi/pt ?id=uiug.30112104082315&view=1up&seq=1.

Fabricant, Nicole. *Fighting to Breathe: Race, Toxicity, and the Rise of Youth Activism in Baltimore*. University of California Press, 2023.

"Facts about Serpentine Rock and Soil Containing Asbestos in California." University of California, Division of Agriculture and Natural Resources. Accessed March 2, 2022. https://anrcatalog.ucanr.edu/pdf/8399.pdf.

Fagone, Jason, and Cynthia Dizikes. "Radioactive Object Found Near Homes at Hunters Point." *San Francisco Chronicle*, September 18, 2018.

———. "Working in a Wasteland." *San Francisco Chronicle*, July 27, 2018.

Fainstein, Susan. *Restructuring the City: The Political Economy of Urban Redevelopment*. Longman, 1984.

"Federal Post-Closure Rule State Regulations: FAQs." Department of Toxic Substances Control. Accessed October 20, 2023. https://dtsc.ca.gov/federal -post-closure-rule-state-faqs/#easy-faq-349212.

Fennell, Catherine. "Are We All Flint?" *limn* 7 (2016). https://limn.it/articles/are -we-all-flint/.

Field, Les, Alan Leventhal, Dolores Sanchez, and Rosemary Cambra. "A Contemporary Ohlone Tribal Revitalization Movement: A Perspective from the Muwekma Costanoan/Ohlone Indians of the San Francisco Bay Area." *California History* 71, no. 3 (1992): 412–431.

"First Year in the Forefront." *Sun-Reporter*, September 13, 1969.

Five Point. "Candlestick Point." Accessed March 2, 2022. www.candlesticksf.com/.

Fredericks, Rosalind. *Garbage Citizenship: Vital Infrastructures of Labor in Dakar, Senegal.* Duke University Press, 2018.

Fredrickson, Leif, Christopher Sellers, Lindsey Dillon, Jennifer Liss Ohayon, Nicholas Shapiro, Marianne Sullivan, Stephen Bocking, et al. "History of US Presidential Assaults on Modern Environmental Health Protection." *American Journal of Public Health* 108, no. S2 (2018): S95–S103.

Fuller, Thomas. "The Loneliness of Being Black in San Francisco." *New York Times,* July 20, 2016.

Galeano, Eduardo. *Open Veins of Latin America: Five Centuries of the Pillage of a Continent.* NYU Press, 1997.

Gandy, Matthew. *Concrete and Clay: Reworking Nature in New York City.* MIT Press, 2003.

Garone, Philip. *The Fall and Rise of Wetlands of California's Great Central Valley.* University of California Press, 2011.

"George Earl on Hunters Point Tenants Union." TV broadcast, December 20, 1966, KRON-TV. Bay Area TV Archive, San Francisco State University.

Gilio-Whitaker, Dina. "As Long as Grass Grows: The Indigenous Fight for Environmental Justice, From Colonization to Standing Rock." Beacon Press, 2019.

Gilmore, Ruth Wilson. "Abolition Geography and the Problem of Innocence." In *Futures of Black Radicalism*, edited by Gaye Theresa Johnson and Alex Lubin. Verso Books, 2017.

———. "Forgotten Places and the Seeds of Grassroots Planning." In *Engaging Contradictions: Theory, Politics, and Methods of Activist Scholarship*, edited by Charles Hale, 31–61. University of California Press, 2008.

———. *Golden Gulag: Prisons, Surplus, Crisis, and Opposition in Globalizing California.* University of California Press, 2007.

Goetz, Edward. *New Deal Ruins: Race, Economic Justice, & Public Housing Policy.* Cornell University Press, 2013.

Goodyear, Charlie. "Protest against PGE Plant." *San Francisco Chronicle,* December 9, 2004.

Gotham, Kevin Fox, and Miriam Greenberg. *Crisis Cities: Disaster and Redevelopment in New York and New Orleans.* Oxford University Press, 2014.

Gould Kenneth A., and Tammy L. Lewis, *Green Gentrification: Urban Sustainability and the Struggle for Environmental Justice.* Routledge, 2016.

Green, Emily. "Infighting Ties Up Millions in Bayview-Hunters Point Grants." *San Francisco Chronicle*, July 25, 2015.

Greenaction for Health and Environmental Justice. *Ticking Time Bomb: Climate Change, Sea Level and Groundwater Rise, Shoreline Contamination, and Environmental Justice in the San Francisco Bay Area.* April 2023. https://greenaction.org/wp-content/uploads/2023/04/Greenaction-Report-April-2023-Shoreline-Contamination-and-Sea-Level-Rise-in-the-San-Francisco-Bay-Area.pdf.

Greenberg, Miriam. "'The Sustainability Edge': Competition, Crisis, and the Rise of Green Urban Branding." In *Sustainability as Myth and Practice in the Global City*, edited by Cindy Isenhour, Gary McDonogh, and Melissa Checker, 105–130. Cambridge University Press, 2015.

———. "What on Earth Is Sustainable? Toward Critical Sustainability Studies." *Boom: A Journal of California* 3, no. 4 (2013): 54–66.

Gregson, Nicky, Helen Watkins, and Melania Calestani. "Inextinguishable Fibres: Demolition and the Vital Materialisms of Asbestos." *Environment and Planning A* 42, no. 5 (2010): 1065–1083.

Gregory, Steven. *Black Corona: Race and the Politics of Place in an Urban Community*. Princeton University Press, 1999.

"Group Demands Better Housing Tonight." *Spokesman*, September 29, 1965.

Gusterson, Hugh. *People of the Bomb: Portraits of America's Nuclear Complex*. University of Minnesota Press, 2004.

Hackel, Steven W. "Land, Labor, and Production: The Colonial Economy of Spanish and Mexican California." In *Contested Eden: California Before the Gold Rush*, edited by Ramón Gutiérrez and Richard Orsi, 111–146. University of California Press, 1998.

Hackworth, Jason. *The Neoliberal City: Governance, Ideology, and Development in American Urbanism*. Cornell University Press, 2019.

Hamusek-McGann, Blossom, Cindy L. Baker, and Mary L. Maniery. *Archeological Inventory and Assessment of Hunters Point Shipyard, San Francisco County, California*. Engineering Field Activity, West, Naval Facilities Engineering Command, 1998.

Harding, Sandra. "'Strong Objectivity': A Response to the New Objectivity Question." *Synthese* 104, no. 3 (1995): 331–349.

Harkewicz, Laura J. "'The Ghost of the Bomb': The Bravo Medical Program, Scientific Uncertainty, and the Legacy of US Cold War Science, 1954–2005." PhD diss., University of California, San Diego, 2010.

Harris, Cheryl. "Of Blackness and Indigeneity: Comments on Jodi A. Byrd's 'Weather with You: Settler Colonialism, Antiblackness, and the Grounded Relationalities of Resistance.'" *Critical Ethnic Studies* 5, nos. 1–2 (2019): 215–228.

Harrison, Jill Lindsey. "Environmental Justice and the State." *Environment and Planning E: Nature and Space* (2022). https://doi-org.oca.ucsc.edu/10.1177/251484862211 3.

———. *From the Inside Out: The Fight for Environmental Justice within Government Agencies*. MIT Press, 2019.

Harrison, Marie. "The Air Up Here." *San Francisco Bayview*, September 17, 2003.

———. "BVHP Mothers Fight for Their Children's Environmental Health." *San Francisco Bayview*, October 20, 2004.

———. "The Taking of Bayview Hunters Point." *San Francisco Bayview*, March 1, 2006.

Hartman, Chester, and Sarah Carnochan. *City for Sale: The Transformation of San Francisco*. University of California Press, 2002.

Hartman, Saidiya. *Lose Your Mother: A Journey Along the Atlantic Slave Route*. Macmillan, 2008.Harvey, David. "From Managerialism to Entrepreneurialism: The Transformation in Urban Governance in Late Capitalism." *Geografiska Annaler: Series B, Human Geography* 71, no. 1 (1989): 3–17.

———. *The New Imperialism*. Oxford University Press, 2005.

Hawthorne, Camilla, and Jovan Scott Lewis. "Black Geographies: Material Praxis of Black Life and Study." In *The Black Geographic Praxis, Resistance, Futurity*, edited by Camilla Hawthorne and Jovan Scott Lewis, 1–24. Duke University Press, 2023.

Henderson, George L. *California and the Fictions of Capital*. Temple University Press, 2003.

Hepler, Lauren. "New Legal Challenge Revives 'Huge War' Over Hunters Point's Toxic Legacy." *San Francisco Chronicle*, June 18, 2021.

Hepler-Smith, Evan. "Molecular Bureaucracy: Toxicological Information and Environmental Protection." *Environmental History* 24 (2019): 534–560.

Herrera, Juan. *Cartographic Memory: Social Movement Activism and the Production of Space*. Duke University Press, 2022.

Hess, David J. *Undone Science: Social Movements, Mobilized Publics, and Industrial Transitions*. MIT Press, 2016.Hird, John. *Superfund: The Political Economy of Risk*. Johns Hopkins University Press, 1994.

Hirsch, Arnold R. *Making the Second Ghetto: Race and Housing in Chicago 1940–1960*. University of Chicago Press, 2009.

Hirsch, Daniel. "EPA to Hunters Point: Pound Sand. No Cleanup as Promised for Superfund Site." *SF Examiner*, October 5, 2022.

Hirsh, Daniel, Taylor Altenbern, Maria Caine, Haakon Williams, and Devyn Gortner. "Hunters Point Shipyard Cleanup Used Outdated and Grossly Non-Protective Cleanup Standards." October 2018. www.committeetobridgethegap.org/wp-content/uploads/2018/10/HuntersPtReport3CleanupStandards.pdf.

"History of the Neighborhood Co-op." *Spokesman*, February 19, 1966.

Holifield, Ryan. "Neoliberalism and Environmental Justice in the United States Environmental Protection Agency: Translating Policy into Managerial Practice in Hazardous Waste Remediation." *Geoforum* 35, no. 3 (2004): 285–297.

Hornbeck, David. "Land Tenure and Rancho Expansion in Alta California, 1784–1846." *Journal of Historical Geography* 4, no. 4 (1978): 371–390.

Horne, Gerald. *Fire This Time: The Watts Uprising and the 1960s*. University of Virginia Press, 1995.

Horowitz, Andy. *Katrina: A History, 1915–2015*. Harvard University Press, 2020.

Horowitz, David Mamaril, and Will Jarrell. "2020 US Census: As San Francisco Grew, the Ethnic Makeup of Its Neighborhood Changed; Here's How." Mission Local, September 21, 2021. https://missionlocal.org/2021/09/census-2020-as-san-francisco-grew-the-ethnic-makeup-of-its-neighborhoods-changed-heres-how/.

Horsman, Reginald. *Race and Manifest Destiny: The Origins of American Racial Anglo-Saxonism.* Harvard University Press, 1981.

Hosseini, Khaled. "Closing Cave of the Winds: A Personal History of Candlestick." *Grantland*, January 9, 2014. https://grantland.com/features/memories-san-francisco-49ers-candlestick-park/.

Hourcle, Laurent R., and Norman H. Guenther. "Institutional Controls for Future Land Use at Active Installation Restoration Program (IRP) Sites." *Remediation Journal* 9, no. 2 (1999): 73–85.

"Housing at Hunters Point: What Will Happen Next?" *Sun-Reporter*, November 17, 1973.

"Housing Authority Faces Another Court Action." *Sun-Reporter*, October 4, 1952.

"Housing Commission Forced to Listen." *Spokesman*, March 19, 1966.

"How Block Clubs Work for Area." *Spokesman*, September 2, 1965.

Hua, Vanessa. "In Bayview, It's Not Business as Usual with Rail Project." *San Francisco Chronicle*, July 20, 2003.

Hula, Richard C. "Changing Priorities and Programs in Toxic Waste Policy: The Emergence of Economic Development as a Policy Goal." *Economic Development Quarterly* 15, no. 2 (2001): 182.

"Hundreds Join Rent Strike." *Spokesman*, November 12, 1966.

"Hunters Point: Community at Crossroads." *Sun-Reporter*, January 15, 1972.

Hunters Point & India Basin Industrial Park Calendar. Joseph L. Alioto Papers, 1958–1977, box 17. San Francisco Public Library, San Francisco.

"Hunters Point Delegation Returns Victorious." *Sun-Reporter* May 16, 1970.

"Hunters Point Dissidents Want Action." Joseph L. Alioto Papers, 1958–1977, folder 11, box 9. San Francisco Public Library, San Francisco.

"Hunters Point Hill Study." Hunters Point Redevelopment, San Francisco Redevelopment Agency Files, folder 24. San Francisco Public Library, San Francisco.

"Hunters Point Housing by and for Hunters Point Residents Phase 1, Five Sites for Moderate Private Housing," October 1968. San Francisco Redevelopment Agency Files. San Francisco Public Library, San Francisco.

"Hunters Point Naval Shipyard: Special Task Force Preliminary Report." John Anderson papers, unfiled. San Francisco Public Library, San Francisco.

"Hunters Point Naval Shipyard Economic Re-Use Study." Mayor's Office of Economic Development, City and County of San Francisco, 1973, John Andersen papers, unfiled. San Francisco Public Library, San Francisco.

"Hunters Point Power Plant Reaches Impasse." *Organized Labor*, July 1, 1996.

Hunters Point Restoration Advisory Board, Reporter's Transcript of Meeting, August 23, 2001. U.S. Navy Public Information Repository, Government Information Center. San Francisco Public Library.

Hunters Point Restoration Advisory Board, Reporter's Transcript of Meeting, September 29, 2001. U.S. Navy Public Information Repository, Government Information Center. San Francisco Public Library.

Hunters Point Restoration Advisory Board, Transcript of Meeting, May 27, 2004. UCSF Chemical Industry Documents. BVHP Community Advocates Collection. Bayview Hunters Point Community Advocates. www.industry documents.ucsf.edu/docs/lxbk0231.

"Hunters Point Shipyard: A Community History." San Francisco Office of Community Investment and Infrastructure, February 1996. Accessed January 15, 2022. https://sfocii.org/sites/default/files/FileCenter/Documents /307-Hunters%20Point%20Shipyard%20A%20Community%20History%20 %20February%201996.pdf.

"Hunters Point Shipyard Closure Protest." TV broadcast, April 26, 1973, KPIX Eyewitness News. Bay Area Television Archives, San Francisco State University.

"Hunters Point Shipyard Residents Reach $6.3M Settlement in Radiation Lawsuit." *CBS SF Bay*, March 25, 2021. https://sanfrancisco.cbslocal.com /2021/03/25/hunters-point-shipyard-residents-radiation-settlement/.

"Hunters Point Toxics—16 Charges Filed against Shipyard." *San Francisco Chronicle*, February 10, 1989.

"Hunters Pointers Want to Stay." *Spokesman*, October 5, 1965.

"Hunters Redevelopment Plan Is Presented." *San Francisco Chronicle*, June 3, 1994.

Hunters View Associates, LLC. "Hunters View Project Description." Accessed October 20, 2023. https://huntersview.info/wp-content/uploads/2014/05 /Full-Project-Description.pdf.

Hunters View Community Partners. "Design for Redevelopment." 2008, 28. https:// huntersview.info/wp-content/uploads/2014/05/Design-for-Development.pdf.

"Hunters View Housing Development: Historic Resources Evaluation." Carey & Co. Inc., July 26, 2001. San Francisco Public Library.

Hurley, Andrew. "From Factory Town to Metropolitan Junkyard: Postindustrial Transitions on the Urban Periphery." *Environmental History* 21, no. 1 (January 2016): 3–29.

Hurtado, Albert L. *Indian Survival on the California frontier.* Yale University Press, 1988.

Igler, David. *Industrial Cowboys: Miller & Lux and the Transformation of the Far West, 1850–1920.* University of California Press, 2005.

Immerwahr, Daniel. *Thinking Small: The United States and the Lure of Community Development.* Harvard University Press, 2015.

Jackson, Christina, and Nikki Jones. "Remember the Fillmore: The Lingering History of Urban Renewal in Black San Francisco." In *Black California Dreamin': The Crises of California's African-American Communities*, edited by Clyde Woods et al. UCSB Center for Black Studies Research, 2012. https://escholarship.org/uc/item/63g6128j.

Jas, Nathalie, and Soraya Boudia, eds. *Toxicants, Health and Regulation since 1945*. Routledge, 2015.Jasanoff, Sheila. "The Songlines of Risk." *Environmental Values* 8, no. 2 (1999): 135–152.

Jatkar, Shrayas, and Jonathan K. London. "From Testimony to Transformation: The Identifying Violations Affecting Neighborhoods (IVAN) Program." UC Davis Center for Regional Change. June 2015. https://regionalchange.ucdavis.edu/report/ivan-program.

Jeffries, Michael P. "Ferguson Must Force Us to Face Anti-Blackness." *Boston Globe*, November 28, 2014.

Jelly-Schapiro, Joshua. "High Tide, Low Ebb." In *Infinite City: A San Francisco Atlas*, edited by Rebecca Solnit. University of California Press, 2010.

Johnson, Charles Spurgeon. *The Negro War Worker in San Francisco*. YWCA in connection with the Race Relations Program of the American Missionary Association, 1945. Bancroft Library, Berkeley.

Johnson, Clarence. "Angry Protestors Seek Shipyard Jobs." *San Francisco Chronicle*, April 26, 1994.

Johnson, Leigh. "The Fearful Symmetry of Arctic Climate Change: Accumulation by Degradation." *Environment and Planning D: Society and Space* 28, no. 5 (2010): 828–847.

Johnson, Marilynn S. *The Second Gold Rush: Oakland and the East Bay in World War II*. University of California Press, 1994.

Johnston, Barbara Rose. *Half-Lives and Half-Truths: Confronting the Radioactive Legacies of the Cold War*. School for Advanced Research, 2007.

Jones, Cynthia Gillian. "A Review of the History of US Radiation Protection Regulations, Recommendations, and Standards." *Health Physics* 88, no. 2 (2005): 105–124.

Jones, Steven. "Battle for Bayview." *San Francisco Bay Guardian*, September 26, 2006.

"Joseph Alioto to Lillian Woods," October 7, 1968. Joseph L. Alioto Papers, 1958–1977, folder 11, box 9. San Francisco Public Library.

"Joseph S. Spinelli to Lex J. Byers," June 1, 1973. John Anderson Papers, unfiled. San Francisco Public Library.

Kang, Helen H. "Looking toward Restorative Justice for Redlined Communities Displaced by Eco-Gentrification." *Michigan Journal of Race and Law* 26 (2021): 23–46.

———. "Respect for Community Narratives of Environmental Injustice: The Dignity Right to Be Heard and Believed." *Widener Law Review* 25 (2019): 219.

Katz, Mitchell H. "Health Programs in Bayview Hunter's Point and Recommendations for Improving the Health of Bayview Hunter's Point Residents." San Francisco Department of Public Health, Office of Policy and Planning, 2006. Accessed March 2, 2022. www.sfdph.org/dph/files/reports/StudiesData /HlthProgsBVHPyRecommends07052006B.pdf.

Kay, Jane. "Activists Vow to Stop Power Ship." *SF Examiner*, July 21, 2000.

———. "Pollution Fears Stir Activists in Hunters Point." *SF Examiner*, February 26, 1996.

Kelley, Robin D. G. *Freedom Dreams: The Black Radical Imagination*. Beacon Press, 2002.

Kilner, Savannah, and Magie Ramírez. "Introduction to Indigenous Geographies of Resurgence." In *Counterpoints: A San Francisco Bay Area Atlas of Displacement and Resistance*, 67–70. PM Press, 2021.

King, John. "Panel Suggests Guidelines for Hunters Point." *San Francisco Chronicle*, March 22, 1993.

Kloc, Ken. *Air Pollution & Environmental Inequity in the San Francisco Bay Area*. Center on Urban Environmental Law, Golden Gate University School of Law, 2011.

Knight, Heather. "Hunters View—Not Sunnydale—Ranks as City's Worst Complex," *SF Gate*, August 29, 2007.

Knuth, Sarah. "Seeing Green in San Francisco: City as Resource Frontier." *Antipode* 48, no. 3 (2016): 626–644.

Knuth, Sarah, Shaina Potts, and Jenny E. Goldstein. "In Value's Shadows: Devaluation as Accumulation Frontier." *Environment and Planning A: Economy and Space* 51, no. 2 (2019): 461–466.

Konisky, David M., ed. *Failed Promises: Evaluating the Federal Government's Response to Environmental Justice*. MIT Press, 2015.

Kramer, Ralph. *Participation of the Poor: Comparative Community Case Studies in the War on Poverty*. Prentice-Hall, 1969.

Krupar, Shiloh. *Hotspotters Report: Military Fables of Toxic Waste*. University of Minnesota Press, 2013.

Kurtz, Hilda. "Acknowledging the Racial State: An Agenda for Environmental Justice Research." *Antipode* 41, no. 4 (2009): 684–704.

Lai, Clement. "The Racial Triangulation of Space: The Case of Urban Renewal in San Francisco's Fillmore District." *Annals of the Association of American Geographers* 102, no. 1 (2012): 151–170.

Lang, Steven, and Julia Rothenberg. "Neoliberal Urbanism, Public Space, and the Greening of the Growth Machine: New York City's High Line park." *Environment and Planning A: Economy and Space* 49, no. 8 (2017): 1743–1761.

Lapp, Rudolph. *Blacks in Gold Rush California*. Yale University Press, 1977.

"Large Crowd Mourns Youth Slain by Police." *Sun-Reporter*, April 12, 1969.

Lefebvre, Henri. *The Production of Space*. Translated by Donald Nicholson-Smith. Blackwell, 1991.

Lelchuk, Ilene. "City Confirms Workers' Charges." *SF Gate*, September 10, 2004. www.sfgate.com/politics/article/City-confirms-workers-charges-SLUG -employees-2726121.php.

Lennar. "Hunters Point Shipyard." March 10, 2005. Wayback Machine, Internet Archive. https://web.archive.org/web/20050310161728/http:/www.hunters pointshipyard.com/index.html.

———. "Hunters Point Shipyard." November 25, 2005. Wayback Machine, Internet Archive. https://web.archive.org/web/20051125181815/http:/www .hunterspointshipyard.com/.

———. "Hunters Point Shipyard." March 16, 2008. Wayback Machine, Internet Archive. https://web.archive.org/web/20080316002128/http:/www.hunters pointcommunity.com/business.html.

———. "Hunters Point Shipyard." June 24, 2015. Wayback Machine, Internet Archive. Accessed March 2, 2022. https://web.archive.org/web/201506 2400 3350/http:/thesfshipyard.com/vision/.

"Lennar Loss Narrower Than Expected." Reuters, September 23, 2008. www.reuters.com/article/us-lennar99/lennar-loss-narrower-than-expected -idUSTRE48M5XV20080923.

Lewis, Jovan Scott. "Black Life Beyond Injury: Relational Repair and the Reparative Conjuncture." *Political Geography* (2023): 1–8.

———. *Violent Utopia: Dispossession and Black Restoration in Tulsa*. Duke University Press, 2022.

Liboiron, Max, Manuel Tironi, and Nerea Calvillo. "Toxic Politics: Acting in a Permanently Polluted World." *Social Studies of Science* 48, no. 3 (2018): 331–349. Lightfoot, Kent G., Lee M. Panich, Tsim D. Schneider, and Sara L. Gonzalez. "European Colonialism and the Anthropocene: A view from the Pacific Coast of North America." *Anthropocene* 4 (2013): 101–115.

"Lillian Woods to Joseph L. Alioto, M. Justin Herman, and Robert B. Pitts," September 5, 1968. Joseph L. Alioto Papers, 1958–1977, folder 11, box 9. San Francisco Public Library, San Francisco.

Lipsitz, George. *The Possessive Investment in Whiteness: How White People Profit from Identity Politics*. Temple University Press, 2006.

Lochhead, Carolyn. "EB-5 Visas Given To Foreign Investors under Fire." *San Francisco Chronicle*, November 2, 2015.

Loyd, Jenna. *Health Rights Are Civil Rights: Peace and Justice Activism in Los Angeles, 1963–1978*. University of Minnesota Press, 2014.

Luby, Edward M., Clayton D. Drescher, and Kent G. Lightfoot. "Shell Mounds and Mounded Landscapes in the San Francisco Bay Area: An Integrated Approach." *Journal of Island & Coastal Archaeology* 1, no. 2 (2006): 191–214.

"M. Justin Herman to Harry D. Ross," July 3, 1961. John F. "Jack" Shelley Papers, 1953–1967, box 2, folder 20. San Francisco Public Library, San

Francisco; and "Housing Trouble Ahead," *San Francisco Examiner*, June 29, 1961.

Maantay, Juliana. "Zoning, Equity, and Public Health." *American Journal of Public Health* 91, no. 7 (2001): 1033.

Madley, Benjamin. *An American Genocide: The United States and the California Indian Catastrophe, 1846–1873*. Yale University Press, 2016.

Maharawal, Manissa. "Protest of Gentrification and Eviction Technologies in San Francisco." *Progressive Planning* 199 (2014): 20–24.

"Major Dates Relating to the Hunters Point Project." John F. "Jack" Shelley Papers, 1953–1967, folder 20, box 2. San Francisco Public Library, San Francisco.

Malin, Stephanie A., and Stacia S. Ryder. "Developing Deeply Intersectional Environmental Justice Scholarship." *Environmental Sociology* 4, no. 1 (2018): 1–7.

Mansfield, Becky. "Particulate Matters: Trump EPA Deregulatory Science, Fossil Fuels, and Racist Regimes of Breathing." *Antipode* 54, no. 4 (2022): 1208–1227.

Markusen, Ann R. *The Rise of the Gunbelt: The Military Remapping of Industrial America*. Oxford University Press, 1991.

Masco, Joseph. "The Age of Fallout." *History of the Present* 5, no. 2 (2015): 137–168.

Massey, Doreen. "Geographies of Responsibility." *Geografiska Annaler: Series B, Human Geography* 86, no. 1 (2004): 5–18.

Massey, Douglas S., and Nancy A. Denton. *American Apartheid: Segregation and the Making of the Underclass*. Routledge, 2019.

Matier & Ross. "Candlestick Park Will Go Out with a Wrecking Ball, Not a Bang." *SFGate*, January 16, 2015. www.sfgate.com/bayarea/matier-ross /article/Candlestick-Park-will-go-out-with-a-wrecking-6019420.php.

Matsuda, Matt K. *Pacific Worlds: A History of Seas, Peoples, and Cultures*. Cambridge University Press, 2012.

"Mayor Alioto on Keeping the Hunters Point Shipyard Open." TV broadcast, June 27, 1973, KPIX Eyewitness News. Bay Area TV Archive, San Francisco State University.

Mays, Kyle. *City of Dispossessions: Indigenous Peoples, African Americans, and the Creation of Modern Detroit*. University of Pennsylvania Press, 2022.

McCormick, Erin. "Bayview's Black Exodus." *San Francisco Chronicle*, January 14, 2008.

———. "Southeast Side Residents Fight Against Power Plants, They Say They Bear the Brunt of Pollution." *San Francisco Chronicle*, September 12, 2003.

McElroy, Erin. "Postsocialism and the Tech Boom 2.0: Techno-Utopics of Racial/Spatial Dispossession." *Social Identities* 24, no. 2 (2018): 206–221.

McEvoy, Arthur. *The Fisherman's Problem: Ecology and Law in the California Fisheries, 1850–1980*. Cambridge University Press, 1986.

McGurty, Eileen. *Transforming Environmentalism: Warren County, PCBs, and the Origins of Environmental Justice*. Rutgers University Press, 2009.

McKittrick, Katherine. *Demonic Grounds: Black Women and the Cartographies of Struggle.* University of Minnesota Press, 2006.

McLaughlin, Malcolm. "The Pied Piper of the Ghetto: Lyndon Johnson, Environmental Justice, and the Politics of Rat Control." *Journal of Urban History* 37, no. 4 (2011): 541–561.

"Men Take Over." *Spokesman,* December 10, 1966.

Merrifield, Andrew. "Place and Space: A Lefebvrian Reconciliation." *Transactions of the Institute of British Geographers* 18, no. 1 (1993): 516–531.

Meyer, Peter B. "Brownfields, Risk-Based Corrective Action, and Local Communities." *Cityscape* 12, no. 3 (2010): 55–69.

Millard, Max. "Power Plant a 'Fantasy,' Say Critics at Public Hearing." *Sun-Reporter,* November 20, 1995.

Miller, Paul T. *The Postwar Struggle for Civil Rights: African Americans in San Francisco, 1945–1975.* Routledge, 2009.

Miller, Richard. *Under the Cloud: The Decades of Nuclear Testing.* The Free Press, 1986.

Misra, Kiran. "Illinois City's Reparations Plan Was Heralded—But Locals Say It's a Cautionary Tale" *Guardian,* August 18, 2021.

Mojadad, Ida, and Laura Waxmann. "Toxic Relationship: The Fraud at Hunters Point." *SF Weekly,* August 22, 2018.

Molanphy, Tom. "'We Feel Abandoned': Bayview Protest Highlights Ongoing Toxic Waste Scandal." 48 Hills, February 14, 2022. https://48hills.org/2022/02/we-feel-abandoned-bayview-protest-highlights-ongoing-toxic-waste-scandal/

Mollenkopf, John. *The Contested City.* Princeton University Press, 1983.

"Monthly Progress Report, Radiation Laboratory, for Period Ending 31 March 1947." General Correspondence, Naval Radiological Defense Laboratory, 1946–1948, Records of Naval Districts and Shore Establishments, Record Group 181. National Archives Branch Depository, San Francisco.

Montoya, María E. *Translating Property: The Maxwell Land Grant and the Conflict over Land in the American West, 1849–1900.* University of California Press, 2002.

Moore, Richard, dir. *Take This Hammer.* KQED, 1964. Bay Area Television Archive, San Francisco State University.

Moore, Shirley Ann Wilson. *To Place Our Deeds: The African American Community in Richmond, California, 1910–1963.* University of California Press, 2000.

Movement for Black Lives. "Reparations Now! Toolkit." Accessed March 2, 2022. https://m4bl.org/wp-content/uploads/2020/05/Reparations-Now-Toolkit-FINAL.pdf.

"Ms. Julia Commer Expresses Needed Ideas." *Spokesman,* January 6, 1966.

Murphy, Michelle. "Alterlife and Decolonial Chemical Relations." *Cultural Anthropology* 32, no. 4 (2017): 494–503.

——. *Sick Building Syndrome and the Problem of Uncertainty: Environmental Politics, Technoscience, and Women Workers*. Duke University Press, 2006.

Nardone, Anthony, Joan A. Casey, Rachel Morello-Frosch, Mahasin Mujahid, John R. Balmes, and Neeta Thakur. "Associations between Historical Residential Redlining and Current Age-Adjusted Rates of Emergency Department Visits Due to Asthma across Eight Cities in California: An Ecological Study." *Lancet Planetary Health* 4, no. 1 (2020): e24–e31.

Nash, Linda. "From Safety to Risk: The Cold War Contexts of American Environmental Policy." *Journal of Policy History* 29, no. 1 (2017): 1–33.

Nelson, Alondra. *Body and Soul: The Black Panther Party and the Fight against Medical Discrimination*. University of Minnesota Press, 2011.

Nelson, Robert K., LaDale Winling, Richard Marciano, Nathan Connolly, et al. "Mapping Inequality." In *American Panorama*, edited by Robert K. Nelson and Edward L. Ayers. Accessed July 7, 2023. https://dsl.richmond.edu /panorama/redlining/#loc=12/37.758/-122.51&city=san-francisco-ca.

"New Anti-Poverty Outpost Plan Proposed." *Spokesman*, September 29, 1965.

"New Gardens," *San Francisco Urban Gardeners Newsletter*, Fall 1988.

"New Health Center for Hunters Point." *Sun-Reporter*, March 13, 1976.

Nixon, Rob. *Slow Violence and the Environmentalism of the Poor*. Harvard University Press, 2011.

Norbert P. Page to John A Sutro. June 29, 1973. John Anderson Papers. San Francisco Public Library, San Francisco.

Noterman, Elsa. "Fugitive Dust: The Indeterminate Trajectories of Urban Development's Present Past." *Annals of the American Association of Geographers* 113, no. 4 (2023): 858.

O'Brien, Tricia. *San Francisco's Bayview Hunters Point*. Arcadia Publishing, 2005.

O'Connor Alice. *Poverty Knowledge: Social Science, Social Policy, and the Poor in Twentieth-Century US History*. Princeton University Press, 2001.

Ohayon, Jennifer Liss. "Addressing Environmental Risks and Mobilizing Democracy? Policy on Public Participation in U.S. Military Superfund Sites." In *Proving Grounds: Militarized Landscapes, Weapons Testing, and the Environmental Impact of US Bases*, edited by Edwin A. Martini, 175–210. University of Washington Press, 2015.

Olmsted, Roger. *Rincon de las Salinas y Potrero Viejo, the Vanished Corner*. Historical Archeological Program, Southeast Treatment Plant, 1978–1979. San Francisco Public Library.

128 Hours: A Report of the Civil Disturbance in the City and County of San Francisco. San Francisco Police Department, 1966. https://archive.org /details/128hoursreportof1968long/.

"Operations Crossroads, 1946." United States, Atmospheric Nuclear Weapons Tests, Nuclear Test Personnel Review. Prepared by the Defense Nuclear Agency as Executive Agency for the Department of Defense, 1984.

"The Outbreak: Moment to Moment." *Spokesman,* October 8, 1966.

Paddock, Richard. "Vision for Transforming Hunters Point Comes before Supervisors." *New York Times,* April 30, 2010.

"Panthers Give Food to Needy." *Sun-Reporter,* October 10, 1970.

Papadopoulos, George. "Chemicals, Ecology, and Reparative Justice." In *Reactivating Elements: Chemistry, Ecology, Practice,* edited by Dimitris Papadopoulos, M. Puig de la Bellacasa, and N. Myers, 34–69. Durham: Duke University Press, 2021.

Papazoglakis, Sarah. "'Feminist, Gun-Toting Abolitionist with a Bankroll': The Black Radical Philanthropy of Mary Ellen Pleasant." *New Global Studies* 12, no. 2 (2018): 235–256.

Pellow, David. "Black Lives Matter as an Environmental Justice Challenge." In *Lessons in Environmental Justice: From Civil Rights to Black Lives Matter to Idle No More,* edited by Michael Mascarenhas, 307–321. Sage, 2021.

Pellow, David Naguib. *What Is Critical Environmental Justice?* John Wiley & Sons, 2017.

Pepin, Elizabeth, and Lewis Watts. *Harlem of the West: The San Francisco Fillmore Jazz Era.* Chronicle Books, 2006.

Perkins, Tracy. *Evolution of a Movement: Four Decades of California Environmental Justice Activism.* University of California Press, 2022.

Pezzullo, Phaedra C. *Toxic Tourism: Rhetorics of Pollution, Travel, and Environmental Justice.* University of Alabama Press, 2009.

Phelan, Sarah. "Air District Fined Lennar Half a Million Last Month." 48 Hills, October 2, 2008. https://sfbgarchive.48hills.org/sfbgarchive/2008/10/02/air -district-fined-lennar-half-million-dollars-last-month/.

———. "Selling Our City to Lennar Corporation." *Race, Poverty & the Environment* 15, no. 1 (2008): 33–35.

"Planning Brd Firm with Herman." *Spokesman,* January 6, 1966.

"'Plant in' at Death Site." *Spokesman,* October 16, 1966.

"Poor Old S.F. Came in Second in Jobs Lost." *San Francisco Chronicle,* January 12, 1989.

"Portrait of a Community Worker: George Earl, the New Breed." *Spokesman,* December 14, 1965.

"Portrait of a Community Worker: Osceola Washington." *Spokesman,* September 17, 1966.

Principles of Environmental Justice. Accessed October 20, 2023. ejnet.org/ej /principles.pdf.

Proctor, Robert N., and Londa Schiebinger. *Agnotology: The Making and Unmaking of Ignorance.* Stanford University Press, 2008.

"Property vs. Human Rights: Otherwise It Was a Great Day for Hunters Point." *Hunters Point-Bayview Community Health Service Newsletter,* August/ September 1969. Bancroft Library, Berkeley.

Public Employees for Environmental Responsibility. "Flint, Environmental Racism, and Racial Capitalism." *Capitalism Nature Socialism* 27, no. 3 (2016): 1–16.

———. "Geographies of Race and Ethnicity II: Environmental Racism, Racial Capitalism and State-Sanctioned Violence." *Progress in Human Geography* 41, no. 4 (2017): 524–533.

———. "Hunters Point Radiation Problems Worsen." May 22, 2018. https://peer .org/hunters-point-radiation-problems-worsen/.

———. "Rethinking Environmental Racism: White Privilege and Urban Development in Southern California." *Annals of the Association of American Geographers* 90, no. 1 (2000): 12–40.

———. "SF-Owned Hunters Point Parcels Soil Testing Falsified." June 4, 2018. https://peer.org/sf-owned-hunters-point-parcels-soil-testing-falsified/.

Pulido, Laura, and Juan De Lara. "Reimagining 'Justice' in Environmental Justice: Radical Ecologies, Decolonial Thought, and the Black Radical Tradition." *Environment and Planning E: Nature and Space* 1, nos. 1–2 (2018): 76–98.

Pulido, Laura, Ellen Kohl, and Nicole-Marie Cotton. "State Regulation and Environmental Justice: The Need for Strategy Reassessment." *Capitalism Nature Socialism* 27, no. 2 (2016): 12–31.

Pulido, Laura, Steve Sidawi, and Robert O. Vos. "An Archaeology of Environmental Racism in Los Angeles." *Urban Geography* 17, no. 5 (1996): 419–439.

Quastel, Noah. "Political Ecologies of Gentrification." *Urban Geography* 30, no. 7 (2009): 694–725.

Radiation Exposure from Pacific Nuclear Tests: Hearing before the House Subcommittee on Oversight and Investigations of the Committee on Natural Resources, 103rd Cong. (February 24, 1994). www.govinfo.gov/content/pkg /CHRG-103hhrg79784/pdf/CHRG-103hhrg79784.pdf.

Radiation Risk Estimates in Normal and Emergency Situations. NATO Security through Science Series. Accessed March 2, 2022. https://link .springer.com/content/pdf/10.1007%2F1-4020-4956-0.pdf.

Ramo, Alan. "Hunters Point: Energy Development Meets Environmental Justice." 5 *Environmental Law News* 28 (Spring 1996): 28–32.

Ranganathan, Malini. "Thinking with Flint: Racial Liberalism and the Roots of an American Water Tragedy." *Capitalism Nature Socialism* 27, no. 3 (2016): 17–33.

Rechtschaffen, Clifford. "Fighting Back against a Power Plant: Some Lessons from the Legal and Organizing Efforts of the Bayview-Hunters Point Community." *Hastings West-Northwest Journal of Environmental Law and Policy* 3 (1995): 407.

Reese, Ashanté, *Black Food Geographies: Race, Self-Reliance, and Food Access in Washington, DC*. UNC Press, 2019.

Reich, Peter L. "Dismantling the Pueblo: Hispanic Municipal Land Rights in California since 1850." *American Journal of Legal History* 45, no. 4 (2001): 353–370.

Reinhold, Robert. "Navy Is Waging a Battle for San Francisco Port." *New York Times*, September 18, 1988.

"Rent Strike." *Spokesman*, March 19, 1966.

Report of the San Francisco Mayor's Task Force on African-American Out-migration. San Francisco, 2009. https://bayviewmagic.org/wp-content/uploads/sites/4/2010/02/AA-OutMigration-TF-1.pdf.

"Request for Approval to Construct Isotope Storage Building" (HRA 418). In *Historical Radiological Assessment*, vol. II, *Use of General Radioactive Materials, 1939–2003, Hunters Point Shipyard*, appendix D. U.S. Department of the Navy, Naval Sea Systems Command, 2004.

"Residents Start Housing Co-op." *Spokesman*, September 2, 1965.

"Residents Stop Eviction." *Spokesman*, March 19, 1966.

Rios, Jodi. *Black Lives and Spatial Matters: Policing Blackness and Practicing Freedom in Suburban St. Louis*. Cornell University Press, 2020.

"Robert Josten to Harvey Rose." FJoseph L. Alioto Papers, 1958–1977, folder 6, box 9. San Francisco Public Library, San Francisco.

Roberts Chris. "SF Housing Site Planned on Former Nuclear Test Site." *SF Gate*, January 12, 2017.

——. "U.S. Navy Dinged for Pushing for Higher Radioactivity Allowances at Hunters Point." SF Curbed, December 19, 2019. https://sf.curbed.com/2019/12/19/21030219/navy-radioactivity-allowances-clean-up-cleanup-hunters-point.

Robertson, Michelle. "New Census Data: San Francisco Getting Richer, More Crowded." *SFGate*, December 26, 2018.

Robichaud, Andrew. *Animal City: The Domestication of America*. Harvard University Press, 2019.

Robinson, Donald. "Can Your City Control an 'Atomic Accident?'" *Los Angeles Times*. May 11, 1958, K10.

Robinson, Cedric. *Black Marxism: The Making of the Black Radical Tradition*. University of North Carolina Press, 2000.

Rocco, Jim, and Lesley Hay Wilson. "The Evolution of Risk-Based Corrective Action." In *Risk-Based Corrective Action and Brownfields Restorations: Proceedings of Sessions of Geo-Congress 98*. American Society of Civil Engineers, 1998.

Romans, Brian. "Learn the Facts about Serpentinite before Its Removed as California's State Rock." KQED, August 5, 2010. www.kqed.org/quest/6714/learn-the-facts-about-serpentinite-before-its-removed-as-californias-state-rock.

Ross, Andrew. "Hunters Point Shipyard Housing Streamlines Its Name." *San Francisco Chronicle*, January 2, 2014.

Roy, Ananya, Stuart Schrader, and Emma Shaw Crane. "'The Anti-Poverty Hoax': Development, Pacification, and the Making of Community in the Global 1960s." *Cities* 44 (2015): 139–145.

Rubenstein, Steve. "S.F. Rally Against Cancer—High Rates in Bayview-Hunters Point." *San Francisco Chronicle*, September 22, 1995.

Rubin, Jasper. *A Negotiated Landscape: The Transformation of San Francisco's Waterfront since 1950.* University of Pittsburgh Press, 2016.

Russell, Emily. "Superfund and Climate Change: Lessons from Hurricane Sandy." *Natural Resources & Environment* 28 (2013): 3–7.

Rydin, Yvonne. "Indicators as a Governmental Technology? The Lessons of Community-Based Sustainability Indicator Projects." *Environment and Planning D: Society and Space* 25, no. 4 (2007): 610–624.

"S.F. Waterfront Losing Another Shipyard—Hundreds of Workers Laid Off." *San Francisco Chronicle*, February 17, 1987.

Safransky, Sara. *The City after Property: Abandonment and Repair in Post-industrial Detroit.* Duke University Press, 2023.

——. "Greening the Urban Frontier: Race, Property, and Resettlement in Detroit." *Geoforum* 56 (2014): 237–248.

Samaha, Albert. "The Dispossessed: Bayview Homeowners Fight Foreclosures." *SF Weekly*, March 2, 2012.

San Francisco African American Reparations Advisory Committee. *Efforts to Support the Preparation of a San Francisco Reparations Plan.* December 2021. https://sf.gov/sites/default/files/2022-03/SF%20AA%20Reparations%20Advisory%20Committee%20-%20December%202021%20Update.pdf.

San Francisco Civil Grand Jury. "Buried Problems and a Buried Process: The Hunters Point Naval Shipyard in a Time of Climate Change." Civil Grand Jury 2021-2022, City and County of San Francisco, June 2022.

San Francisco Department of City Planning. *South Bayshore, 1970 Census: Population and Housing Summary and Analysis.* February 1973. Government Services Desk, San Francisco Public Library.

"San Francisco Healthy Homes Project: Community Health Status Assessment." San Francisco Department of the Environment, City and Country of San Francisco, and San Francisco Department of Public Health. Accessed March 2, 2022. https://sfenvironment.org/sites/default/files/fliers/files/sfe_ej_sfhh_community_health_status_assessment.pdf.

San Francisco Municipal Transportation Agency. "Bayview Community-Based Transportation Plan." Accessed March 3, 2022.www.sfmta.com/projects/bayview-community-based-transportation-plan.

San Francisco Office of the Assessor. *Annual Reports.* 2006–2022. Accessed October 21, 2023. sfassessor.org/news-information/annual-reports.

San Francisco Office of Community Investment and Infrastructure. "Candlestick Point: Design for Development." Accessed October 21, 2023. https://

sfocii.org/sites/default/files/inline-files/2.%20CANDLESTICK%20POINT
%20DESIGN%20FOR%20DEVELOPMENT_Amended%202019.pdf.
———. "Hunters Point Shipyard/Candlestick Point Disposition and Development Agreement, Exhibit K: Sustainability Plan." San Francisco, 2018.
https://sfocii.org/projects/hunters-point-shipyard-candlestick-point-2
/document-library.
San Francisco Planning Department. *Addendum 3 to Environmental Impact
Report.* September 9, 2014. https://sfocii.org/sites/default/files/inline-files
/Addendum%203%20to%20CP-HPS2%20EIR%202014%200919.pdf.
San Francisco Redevelopment Agency. "Draft Executive Summary Regarding
the Environmental Remediation of the Hunters Point Shipyard." April 2010
.www.sfdph.org/dph/hc/HCAgen/HCAgen2010/files406012010/Attach3
CleanupExecSum.pdf.
———. "Memorandum." September 16, 2008. Wayback Machine, Internet
Archive, https://sfocii.org/ftp/archive/sfarchive.org/indexf1c4.html?dept=
1051&sub=&dtype=3456&year=7344&file=88509.
Sandlos, John, and Arn Keeling. "Zombie Mines and the (Over) burden of
History." *Solutions Journal* 4, no. 3 (2013): 1–4.
Sangyun, Lee, and Paul Mohai. "Environmental Justice Implications of
Brownfield Redevelopment in the United States." *Society & Natural
Resources* 25, no. 6 (2012): 602–609.
Santos, Daniel, Ashley Calvillo, Stephanie Cefalu, María José Ospina Salcedo,
Brea Violett, Helen Kang, and Robert Mullaney. "Concrete Production and
the Regulatory Role of the Bay Area Air Quality Management District."
Golden Gate University Environmental Law and Justice Clinic. May 2020.
Saxton, Alexander. *The Indispensable Enemy: Labor and the Anti-Chinese
Movement in California.* University of California Press, 1995.
Schwartz, Alicia. "Proposition F: A Fight for the Heart and Soul of San Fran-
cisco." *SF Bayview*, July 1, 2008.
Scott, Mel. *The San Francisco Bay Area: A Metropolis in Perspective.* University
of California Press, 1985.
Seale, Bobby. "Black Panthers and Hunters Point." *Black Panther*, July 20, 1967.
Self, Robert. *American Babylon: Race and the Struggle for Postwar Oakland.*
Princeton University Press, 2005.
Sellers, Christopher, Lindsey Dillon, Jennifer Liss Ohayon, Nick Shapiro,
Marianne Sullivan, Chris Amoss, Stephen Bocking, et al. *The EPA under
Siege: Trump's Assault in History and Testimony.* The Environmental Data
and Governance Initiative, June 2017. https://100days.envirodatagov.org/epa
-under-siege/.
Selna, Robert. "Campaign 2006/Bayview-Hunters Point." *San Francisco
Chronicle*, September 29, 2006.
———. "Lennar Corp Dominates Redevelopment in SF." *SF Gate*, April 10, 2007.

———. "News/Politics." *San Francisco Chronicle*, September 29, 2006.

SF Parks Alliance. "Blue Greenway." Accessed October 23, 2023. https://san franciscoparksalliance.org/projects/blue-greenway/.

Shange, Savannah. *Progressive Dystopia: Abolition, Antiblackness, and Schooling in San Francisco*. Duke University Press, 2019.

Shapiro, Nicholas. "Attuning to the Chemosphere: Domestic Formaldehyde, Bodily Reasoning, and the Chemical Sublime." *Cultural Anthropology* 30, no. 3 (2015): 368–393.

Sharpe, Christina. *In the Wake: On Blackness and Being*. Duke University Press, 2016.

"Shipyard Worker's Committee Against Discrimination (SWCAD)." Raymond F. Thompson Papers 1941–1988, box 24. Meiklejohn Civil Liberties Institute Collections, Bancroft Library, UC Berkeley.

"Should Do Something for Neighborhood." *Spokesman*, September 29, 1965.

Simon, Suzanne E. "Editor's Perspective: Trends in Hazardous Site Cleanup and the Remediation Market in the United States." *Remediation* (Spring 2002): 1–4.

Simons, Robert A. *Turning Brownfields into Greenbacks: Developing and Financing Environmentally Contaminated Real Estate*. Urban Land Institute, 1998.

Smiley, Lauren. "The Man Who Cried Dust." *SF Weekly*, July 1, 2009.

Smith, Neil. *The New Urban Frontier: Gentrification and the Revanchist City*. Psychology Press, 1996.

Smith, Stacey. "Remaking Slavery in a Free State: Masters and Slaves in Gold Rush California." *Pacific Historical Review* 80, no. 1 (2011): 28–63.

Solis, Miriam. "Conditions and Consequences of ELULU Improvement: Environmental Justice Lessons from San Francisco, CA." *Journal of Planning Education and Research* (2020): 1–13.

———. "Engineering Justice: Cities, Race, and 21st Century Wastewater Infrastructure." PhD diss. University of California, 2018.

Sorenson, David. *Shutting Down the Cold War: The Politics of Military Base Closure*. St. Martin's Press, 1998.

"The Southern Suburbs: Building Booms at South San Francisco and the Potrero." *San Francisco Call*, May 9, 1895.

State of California Reparations Task Force Meeting. Witness Panel on Environmental Racism, October 2021. https://oag.ca.gov/ab3121/meetings/102021.

"Statement by Secretary of Defense Eliot L. Richardson on Base Realignments." Joseph L. Alioto Papers, 1958–1977, box 4, folder 17. San Francisco Public Library, San Francisco.

"Statement of Stanley R. Larsen, Lt. Gen., U.S. Army, Asst. Deputy for Development, City of SF." John F. "Jack" Shelley Papers, 1953–1967, box 17. San Francisco Public Library, San Francisco.

"A Stone Dry Dock." *Daily Alta California*, October 6, 1864.

"Suggests a New Chinatown." *San Francisco Call*, May 12, 1906.

Sumchai, Ahimsa Porter. "HP Biomonitoring—Promising HOPE for Hunters Point." *San Francisco Bay View*, March 2, 2022.

———. "The RABblerousers!" *San Francisco Bay View*, June 30, 2009.

Summers, Brandi Thompson. *Black in Place: The Spatial Aesthetics of Race in a Post-Chocolate City*. UNC Press Books, 2019.

Sze, Julie. *Environmental Justice in a Moment of Danger*. University of California Press, 2020.

———. *Fantasy Islands: Chinese Fears and Ecological Dreams in an Age of Climate Crisis*. University of California Press, 2015.

Táíwò, Olúfhemi O. *Reconsidering Reparations*. Oxford University Press, 2022.

Takaki, Ronald T. *Iron Cages: Race and Culture in 19th-Century America*. Oxford University Press, 1979.

TallBear, Kim. "Standing with and Speaking as Faith: A Feminist-Indigenous Approach to Inquiry." *Journal of Research Practice* 10, no. 2 (2014): 1–7.

Taylor, Keeanga-Yamahtta. *Race for Profit: How Banks and the Real Estate Industry Undermined Black Homeownership*. UNC Press Books, 2019.

Thompson, Ki'Amber. "Toward a World Where We Can Breathe: Abolitionist Environmental Justice Praxis." *Annals of the American Association of Geographers* 113, no. 7 (2023): 1699–1710.

Thompson, Willie. "Wake Up to the Redevelopment Game!" *Spokesman*, January 7, 1967.

"To Be or Not to Be: A Community Hospital for Southeast S.F." *Hunters Point-Bayview Community Heath Service Newsletter*, December 1969/ January 1970. Bancroft Library, Berkeley.

Tompkins, Christien. "There's No Such Thing as a Bad Teacher: Reconfiguring Race and Talent in Post-Katrina Charter Schools." *Souls* 17, nos. 3–4 (2015): 211–230.

"Topics for Investigation by Radiation Laboratory." U.S. Naval Radiological Defense Laboratory, General Correspondence 1946–48, February 6, 1947. Records of Naval Districts and Shore Establishments, Record Group 181, National Archives Branch Depository, San Francisco.

"Toxic Dumping Judgement Cut to $115,000." *San Francisco Chronicle*, July 1, 1995.

Toxic Wastes and Race in the United States: A National Report on the Racial and Socio-Economic Characteristics of Communities with Hazardous Waste Sites. United Church of Christ Commission on Racial Justice, 1987.

"Transcript of the Proceedings in Case No. 30, Rincon de Salinas y Potrero Viejo, Heirs of José Cornelius Bernal vs. The United States." 5 ND, Documents Pertaining to the Adjudication of Private Land Claims in California, circa 1852–1892. Bancroft Library, Berkeley.

U.S. Department of the Navy, Base Realignment and Closure. "Draft Record of Decision for Parcel E-2, Hunters Point Naval Shipyard." March 2012. www

.bracpmo.navy.mil/brac_bases/california/former_shipyard_hunters_point
/public_notices/cercla_record_of_decision_for_parcel_e2.html.

———. "Final Amended Parcel B Record of Decision." Accessed January 15,
2022. https://semspub.epa.gov/work/09/100002083.pdf.

———. "Hunters Point Naval Shipyard, Final Historical Radiological Assessment: History of the Use of General Radioactive Materials, 1949–2003."
Accessed October 18, 2023. https://media.defense.gov/2022/Mar/02/2002
948270/-1/-1/0/HPS_200408_HRA.PDF.

———. "Proposal to Dissolve the Hunters Point Restoration Advisory Board."
September 1, 2009. Copy in possession of author.

U.S. Environmental Protection Agency. "Brownfields." Accessed October 12,
2023. www.epa.gov/brownfields.

———. "Calculating Preliminary Remediation Goals." Accessed January 15,
2022. www.epa.gov/risk/calculating-preliminary-remediation-goals-prgs.

———. "Hunters Point Naval Shipyard Superfund Site Profile." Accessed
October 22, 2023, https://cumulis.epa.gov/supercpad/cursites/csitinfo.cfm
?id=0902722.

———. "Naturally Occurring Asbestos: Approaches for Reducing Exposure."
Office of Superfund Remediation and Technology Innovation, 2008.
https://archive.epa.gov/region9/toxic/web/html/basic.html.

———. "Restoration Advisory Board (RAB) Implementation Guidelines."
September 27, 1994. www.epa.gov/fedfac/restoration-advisory-board-rab
-implementation-guidelines.

———. "Superfund Cleanup Process." Accessed October 22, 2023. https://www
.epa.gov/superfund/superfund-cleanup-process.

———. Technical Assistance Grant (TAG) Program. www.epa.gov/superfund
/technical-assistance-grant-tag-program.

U.S. Green Building Council. "The Shipyard/Candlestick Point." Accessed
October 22, 2023. www.usgbc.org/projects/shipyardcandlestick-point.

U.S. Government Accountability Office. *Siting of Hazardous Waste Landfills
and Their Correlation with Racial and Economic Status of Surrounding
Communities.* June 1, 1983. gao.gov/products/rced-83-168.

"Untitled Document." NAACP Region 1 Records, 1942–1986, folder 13, box 24.
Bancroft Library, Berkeley.

"Urban Land Institute Panel Analysis: Hunters Point Naval Shipyard at San
Francisco." John Andersen papers, unfiled. San Francisco Public Library,
San Francisco.

"US NRDL History 1946–1955" (HRA-587). In *Historical Radiological Assess-
ment*, vol. II, *Use of General Radioactive Materials, 1939–2003, Hunters
Point Shipyard*, appendix D. U.S. Department of the Navy, Naval Sea
Systems Command, 2004.

Vasudevan, Pavithra. "An Intimate Inventory of Race and Waste." *Antipode* 53,
no. 3 (2021): 770–790.

Vasudevan, Pavithra, and Sara Smith. "The Domestic Geopolitics of Racial Capitalism." *Environment and Planning C: Politics and Space* 38, nos. 7–8 (2020): 1160–1179.

Von Hoffman, Alexander. "Calling upon the Genius of Private Enterprise: The Housing and Urban Development Act of 1968 and the Liberal Turn to Public-Private Partnerships." *Studies in American Political Development* 27, no. 2 (2013): 165–194.

Voyles, Traci Brynne. *Wastelanding: Legacies of Uranium Mining in Navajo Country.* University of Minnesota Press, 2015.

Wachsmuth, David, and Hillary Angelo. "Green and Gray: New Ideologies of Nature in Urban Sustainability Policy." *Annals of the American Association of Geographers* 108, no. 4 (2018): 1038–1056.

Walker, J. Samuel. *Permissible Dose: A History of Radiation Protection in the Twentieth Century.* University of California Press, 2000.

Walker, Richard. "Industry Builds the City: The Suburbanization of Manufacturing in the San Francisco Bay Area, 1850–1940." *Journal of Historical Geography* 27, no. 1 (2001): 36–57.

———. "Landscape and City Life: Four Ecologies of Residence in the San Francisco Bay Area." *Ecumene* 2, no. 1 (1995): 33–64.

———. *Pictures of a Gone City: Tech and the Dark Side of Prosperity in the San Francisco Bay Area.* PM Press, 2018.

Wanzer-Serrano, Darrel. *The New York Young Lords and the Struggle for Liberation.* Temple University Press, 2015.

Washburn, Patrick. *The African American Newspaper: Voice of Freedom.* Northwestern University Press, 2006.

———. "The Pittsburgh Courier's Double V Campaign in 1942." *American Journalism* 3, no. 2 (1986): 73–86.

Washburn, Stephen, and Kristin Edelmann. "Development of Risk-Based Remediation Strategies." In *Risk-Based Corrective Action and Brownfields Restorations: Proceedings of Sessions of Geo-Congress 98.* American Society of Civil Engineers, 1998.

"Washington Notes." *Los Angeles Times,* September 24, 1980, B26.

Waxmann, Laura. "Hunters Point Shipyard Homeowners Announce Lawsuit Against Developer, Say 'We Were Duped.'" *San Francisco Examiner,* July 25, 2018.

Weisgall, Jonathan. *Operation Crossroads: The Atomic Tests at Bikini Atoll.* Naval Institute Press, 1994.

———. "The Nuclear Nomads of Bikini." *Foreign Policy* 39 (1980): 74–98.

Weiss, Marc A. "The Real Estate Industry and the Politics of Zoning in San Francisco, 1914–1928." *Planning Perspectives* 3, no. 3 (1988): 311–324.

Welsome, Eileen. *The Plutonium Files: America's Secret Medical Experiments in the Cold War.* Delta, 2010.

Wernstedt, Kris, and Robert Hersh. "Through a Lens Darkly—Superfund Spectacles on Public Participation at Brownfield Sites." *Risk: Health, Safety & Environment* 9, no. 2 (Spring 1998): 153–174.

Wernstedt, Kris, Robert Hersh, and Katherine Probst. "Basing Superfund Cleanups on Future Land Uses: Promising Remedy or Dubious Nostrum?" Resources for the Future, Discussion Paper 98-03. October 1997. rff.org /publications/working-papers/basing-superfund-cleanups-on-future-land -uses-promising-remedy-or-dubious-nostrum/.

Wheeler, Mary E. "Empires in Conflict and Cooperation: The" Bostonians" and the Russian-American Company." *Pacific Historical Review* 40, no. 4 (1971): 419–441.

Whiting, Sam. "S.F. Is about to Break Ground on the Most Expensive Park in City History." *San Francisco Chronicle*, June 21, 2021.

Williams, Rhonda Y. *The Politics of Public Housing: Black Women's Struggles against Urban Inequality.* Oxford University Press, 2004.

Wirt, Frederick. *Power in the City: Decision Making in San Francisco.* University of California Press, 1974.

Wolfrom, Jessica. "Bayview Residents May Soon Find Relief from Harmful Industrial Dust." *SF Examiner*, March 14, 2022.

———. "Fresh Concerns Raised about Cleanup of Hunters Point Shipyard." *SF Examiner*, April 7, 2023.

Wong, Julia Carrie. "The Bay Area Roots of Black Lives Matter." *SF Weekly*, November 11, 2015.

Woods, Clyde. *Development Arrested: The Blues and Plantation Power in the Mississippi Delta.* Verso Books, 2017.

Wynter, Sylvia. "1492: A New World View." In *Race, Discourse, and the Origin of the Americas: A New World View*, edited by Vera Lawrence Hyatt and Rex Nettleford, 5–57. Smithsonian Institution Press, 1995.

Yesko, Parker, and Steven T. Jones. "Community Awaits Benefits as Lennar Finally Breaks Ground in Hunters Point." 48 Hills. July 1, 2013. https:// sfbgarchive.48hills.org/sfbgarchive/2013/07/01/community-awaits-benefits -lennar-finally-breaks-ground-hunters-point/.

Zoellner, Tom. "Shipyard Fire Poses No Risk, Officials Say." *San Francisco Chronicle*, September 13, 2000.

Index

Potrero Hill, 53, 67, 121

power plants: long-term exposure to toxins and, 142–43; protests against, 3, 5, 117, 120–23, 126–27, 141, 147; redeveloping contaminated sites, 3; resident concerns about, 44, 122; urban geography and, 6*map*, 26, 111

Principles of Environmental Justice, The (1991), 15

Proposition F, 101

Public Employees for Environmental Responsibility, 20

public housing: community gardens at, 18–19; evictions from, 50–51; Hunters Point redevelopment project, 60–66; Hunters Point rent strike, 53–60, 54*fig*; neoliberal transformation of, 113; racialized urban development, postwar, 41–45; SFHA neglect of, 50–60; West Point/Hunters View, 10–11; West Point/Hunters View, HOPE SF redevelopment program, 111–17, 115*fig*

public transportation, 7–8, 114, 154n22

Puerto Rican Young Lords, 56

Pulido, Laura, 15, 16, 25

Quesada Gardens Initiative, 19, 78

RAB. *See* Hunters Point Shipyard Restoration Advisory Board (RAB)

race and racism: black aesthetic emplacement, 142; Double V campaign, 35, 37, 71, 72*fig*; employment opportunities and, 34–38; health care and, 67–68; housing segregation and, 32–33, 35, 37–38, 41–45, 137–38, 149–50, 159n46, 162n86; killing of Matthew Johnson, 46–48, 48*fig*, 49*fig*; power plant protests and environmental racism, 121–23; protests for fair hiring practices, 66–67; racial injustice, Bayview-Hunters Point neighborhood and, 1–6, 5*map*, 6*map*; racialized accumulation of toxicity, 142–44; racialized urban development, postwar, 41–45; racializing surveillance, 138; racial logics of property values, 138; redlining, 32–33, 149–50, 159n46; reparations and environmental justice, 147–52; role of state in racialized environmental harms, 15–16, 150–52; San Francisco population data, 4; *Toxic Wastes and Race in the United States*, 14; U.S. GAO report on race and toxic landfills, 14; wastelanding and, 25–26, 43–45, 163n101;

zoning laws and, 32. *See also* environmental racism, racial capitalism, wastelanding

racial capitalism, 25; garbage, political geography of, 57; industrialization of southeast San Francisco, 31–33, 142–43; Parcel E-2 remediation as, 107; zoning laws and, 32. *See also* wastelanding

racializing surveillance, 138

radioactive waste, 39–40, 160nn73–74, 161n78, 161–62n80; background levels, 172n32; Hunters Point Restoration Advisory Board (RAB) about, 93–96; Hunters Point Shipyard and, 76; Parcel E-2, 85–86; Parcel E-2, protesting landfill cap on, 105–10; sea-level rise, concerns about, 87–88, 140, 173n43; Treasure Island, radioactive spill, 95–96

Radiological Affairs Support Office (RASO), U.S. Navy, 94–96

Rancho rincón de las salinas y potrero viejo, 29

Ralston, William, 31

Ranganathan, Malini, 15, 150–52

Ratcliff, Willie, 108

Rat Extermination and Control Bill, 51

Reagan, Ronald, 81, 171n17

real estate, wastelanding and capital accumulation, 45, 163n101

redevelopment dust, 111–17, 115*fig*; Bayview Hill and demolition of Candlestick Park, 127–34; in Detroit, 116; Hunters Point shipyard demolition, 123–27; India Basin project, 138–42; organizing against air pollution, 117–20; in Philadelphia, 116; power plant protests, 120–23; racialized accumulation of toxicity, 142–44; rebranding redevelopment, 134–38

Redevelopment Project Area Committee (PAC), 77

redlining, 32–33, 149–50, 159n46

regimes of perceptibility, 132

Regional Water Quality Control Board (RWCQB), California, 85–86

regulatory oversight: of Parcel E-2, 85–87; of Superfund sites, 85

remediation: CERCLA, cleanup standards in, 83; debating meaning of cleanup, 87–90; engineering controls, 85; of former military bases, 77, 79–80; institutional controls, 85; local jobs created from, 92–93; Parcel E-2, 84–86; Parcel E-2, protesting landfill cap on, 105–10; as a provisional accomplishment, 88–89;

Founded in 1893,
UNIVERSITY OF CALIFORNIA PRESS
publishes bold, progressive books and journals
on topics in the arts, humanities, social sciences,
and natural sciences—with a focus on social
justice issues—that inspire thought and action
among readers worldwide.

The UC PRESS FOUNDATION
raises funds to uphold the press's vital role
as an independent, nonprofit publisher, and
receives philanthropic support from a wide
range of individuals and institutions—and from
committed readers like you. To learn more, visit
ucpress.edu/supportus.